Cyborgs, Sexuality, and the Undead

The Body in Mexican and Brazilian Speculative Fiction

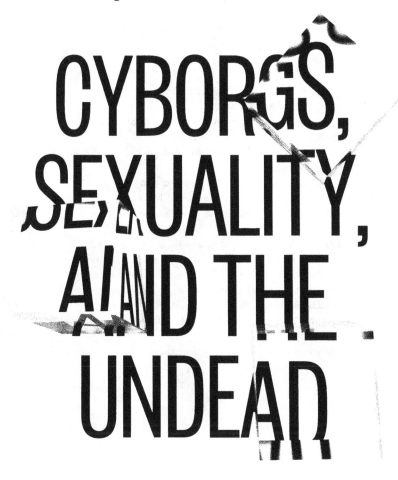

CYBORGS, SEXUALITY, AND THE UNDEAD

M. ELIZABETH GINWAY

VANDERBILT UNIVERSITY PRESS
Nashville, Tennessee

Library of Congress Cataloging-in-Publication Data
Names: Ginway, M. Elizabeth, author.
Title: Cyborgs, sexuality, and the undead : the body in Mexican and
   Brazilian speculative fiction / M. Elizabeth Ginway.
Description: Nashville : Vanderbilt University Press, [2020] | Includes
   bibliographical references and index.
Identifiers: LCCN 2020027459 (print) | LCCN 2020027460 (ebook) | ISBN
9780826501172 (paperback) | ISBN 9780826501189   (hardcover)
 | ISBN 9780826501196 (epub) | ISBN   9780826501202 (pdf)

Subjects: LCSH: Speculative fiction, Mexican—History and criticism. |
   Speculative fiction, Brazilian—History and criticism. | Human body in
   literature. | Monsters in literature. | Gender identity in literature. |
   Comparative literature—Brazilian and Mexican. | Comparative
   literature—Mexican and Brazilian.
Classification: LCC PQ7207.S68 G56 2020 (print) | LCC PQ7207.S68 (ebook)
   | DDC 863/.087609972—dc23
LC record available at https://lccn.loc.gov/2020027459
LC ebook record available at https://lccn.loc.gov/2020027460

In memory of my sister, Jennifer Whitlock Ginway Mistrano (1967–2014)

CONTENTS

Acknowledgments

INTRODUCTION: Rereading the Body in the Speculative
Fiction of Mexico and Brazil   I

1. Gendered Cyborgs: Mechanical, Industrial, and Digital   26

2. The Baroque Ethos, *Antropofagia*, and Queer Sexualities   70

3. Trauma Zombies, Consumer Zombies, and Political Zombies   107

4. Vampires: Immunity and Resistance   136

   Afterword   167

   Notes   169
   Bibliography   197
   Index   221

ACKNOWLEDGMENTS

Writing a book at age sixty is quite different from writing one at age forty, mostly because of the shift in perspective brought about by time and its effects on mind, body, and spirit. I would like to give thanks to those who have helped me along the way. I began to venture down the road of Spanish American science fiction thanks to Andrea L. Bell and Yolanda Molina Gavilán, whose 2003 *Cosmos Latinos*, an anthology of Latin American and Spanish texts in English, inspired me to teach my first class on Latin American SF in English. I ventured further, attending two key Spanish American science fiction symposia during 2010 and 2011. The first was in Santiago, Chile, where l heard talks by J. Andrew Brown, Edmundo Paz Soldán, Alberto Fuguet, Jorge Baradit, and Mike Wilson. The second was held in Tijuana, Mexico, where I had the privilege of meeting Gabriel Trujillo Muñoz and speaking at length with Pepe Rojo, Deyanira Torres, Bernardo Fernández (Bef), Horacio Porcayo, Bruce Sterling, Chris Brown, and Miguel Ángel Fernández Delgado (MAF). Later, MAF—who was extraordinarily generous with his time and knowledge—arranged for me to meet with SF fans in Mexico City, and Bef invited me to a gathering of several writers, including José Luis Zárate, Karen Chacek, Edgar Omar Avilés, Ricardo Guzmán Wolffer, J. M. Rodolfo, and Alberto Chimal. Since 2010, as part of the UF study abroad program, I have traveled to Brazil five times, enabling me to maintain my contacts with science fiction writers Gerson Lodi-Ribeiro and Roberto de Sousa Causo. I also wish to recognize the work of Marcello Simão Branco and César Silva and their publication *Anuário brasileiro de literatura fantástica*, which also allowed me to keep a finger on the pulse of Brazilian science fiction. I would also like to thank Alfredo Suppia for inviting me to UNICAMP in 2016 and for being a supportive colleague during his Fulbright here at the University of Florida 2019–2020. During this year of pandemic, I extend thanks to Brazilian conference organizers Ana Rüsche and Naiara Araújo for inviting me to speak at their virtual events, where I met Bruno Anselmi Matangrano and Alexander Meireles da Silva, albeit virtually.

Closer to home, many colleagues and communities have supported my work, beginning with science fiction scholars Rachel Haywood Ferreira and J. Andrew Brown. Wider communities include frequent attendees of the International Conference on the Fantastic and the Arts, including Dale Knickerbocker, Juan Carlos Toledano, Suparno Banerjee, Pawel Frelik, and Amy Ransom, as well as the editors of *Science Fiction Studies*—Veronica Hollinger, Joan Gordon, and Art Evans. I also wish to thank *Extrapolation* editor and organizer of the Eaton Conference Sherryl Vint for her support. At the University of Florida, Jennifer Rea and Terry Harpold have been especially supportive of international SF events. I thank Emily Hind, my colleague in Spanish and Portuguese Studies, for putting me into contact with Carmen Boullosa, Karen Chacek, and Cristina de la Garza. I also wish to thank Emanuelle Oliveira, David Dalton, Giovanna Rivero, James Krause, and Christopher Lewis for their interest in science fiction and their collaborations, both past and future.

I am grateful to the anonymous readers and the editors of Vanderbilt University Press for their suggestions and dedication to improving the manuscript and getting it into readable form. I also recognize the University of Florida, whose sabbatical program, together with the College of Liberal Arts and Sciences' Scholarship Enhancement grants, offered release time and financial support. I also am grateful for the support of my chair, Gillian Lord. I also acknowledge, with thanks, the subvention grant I received from the University of Florida College of Liberal Arts and Sciences and Center for the

Humanities and the Public Sphere (Rothman Endowment). I thank my aunt Jane Long and her daughter Marti as well as my son Matt McAllen for their support and curiosity about this book. Finally, I am especially thankful for the unwavering support and patience of my husband, David Pharies, who saw me through the ups and downs of this lengthy process, reading many drafts and helping me see the project through to the end.

Portions of Chapter 2 appear in "Resistant Female Cyborgs in Brazil," in "Ibero-American Homage to Mary Shelley," special issue, *Alambique: Revista académica de ciencia ficción y fantasía* 7, no. 1 (2020): Article 5. Portions of Chapter 3 were published in "Transgendering in Brazilian Speculative Fiction from Machado de Assis to the Present," *Luso-Brazilian Review* 47, no. 1 (2010): 40–60, and part of Chapter 4 appears in "Eating the Past: Proto-Zombies in Brazilian Fiction 1900–1955," in "The Transatlantic Unidead: Zombies in Hispanic and Luso-Brazilian Literatures and Cultures," ed. David Dalton and Sara Anne Potter, special issue, *Alambique: Revista académica de ciencia ficción y fantasía* 6, no. 1 (2018): Article 7. Thanks to the editors of the journals for permission to reprint the portions of the articles cited above.

Rereading the Body in the Speculative Fiction of Mexico and Brazil

This study builds on my 2004 monograph *Brazilian Science Fiction: Cultural Myths and Nationhood in the Land of the Future*. Specifically, it focuses on a single theme—the body—rather than on a set of icons or subgenres of science fiction. It extends the generic perspective beyond science fiction to the more general category of speculative fiction, which includes fantasy and horror, and it broadens the literary base to include works originating in Mexico as well as Brazil.

The Body in Mexico and Brazil

While the emphasis on science in the Anglo-American science fiction tradition often accentuates the generalized Western binary between mind and body, Latin American science fiction often appears to find a more nuanced middle ground, where cultural traditions resist the idea of a mind without a body. This explains why the body, especially in its ambiguous state, is the ideal locus for the study of speculative fiction in Mexico and Brazil. The cyborg exists as an amalgam of the cybernetic or mechanical and the organic or biological, relating to the social body in terms of class and race. Nontraditional sexualities represent a state between or beyond the genders, thus serving as an exemplary vehicle for examining gender and sexuality in the broadest sense. Zombies and vampires are the ultimate expression of an intermediary state, because as the living dead and the undead they embody the paradox of a present haunted by an embodied past.

The tales of embodiment presented here spark the imagination and allow readers to gain insight into these two complex cultures—Mexican and Brazilian—and their historic struggles. In these two late-modernizing societies, a diachronic, interpretive overview of the treatment of the body in the counternarratives of speculative fiction is an effective way to understand the evolution of and, more especially, the resistance to modernization in these societies.

My thesis is that bodily transformation is used in these stories to show human resilience in the face of national and global inequalities arising from a past often characterized by authoritarian rule. Despite the official rhetoric of *mestizaje/mestiçagem*, there is a lurking sense that certain bodies continue to be identified as "impolitic," as targets of "thanatopolitics" or "necropolitics," that is, the state's algorithm of race and power used to "regulate the distribution of death" (Mbembe 17).[1] In Mexico and Brazil, where political corruption and the spread of transnational organized crime have eroded any sense of political agency, consensus, and citizenship, the body and its transformations stand as a reminder, beyond sterile abstractions and systemic thinking, of resistance by those who have survived and continue to challenge structures of power and control. This is mirrored in Mexican and Brazilian works of speculative fiction.

The body is a reminder of our physical and social presence in the world, a marker of gender and race. It is also part of a social body, of a collective society with values, traditions, and myths. In countries such as Mexico and Brazil—as in Latin America as a whole—the body plays a central role in religious and other popular collective manifestations, both cultural and political.[2] The metaphorical use of the body is a commonplace in socio-political discourse, easily understood and, consequently, mythified and naturalized. The body of the king, the colonized body of the conquered or enslaved, the body of Christ and of other gods, and the body of the sacred mother are part of the cultural imaginaries of past and present.[3] As countries with colonial histories dating from the sixteenth century, Mexico and Brazil recapitulate the weight and power of the embodied past of conquest and enslavement that led to the formation of these two racially diverse, unique countries, which today are among the largest and most industrialized of Latin America.[4]

In this study, the body is examined from two perspectives: first, in the light of biopolitics, that is, Foucault's conception of the body as an entity controlled and disciplined by the state;[5] and second, as a variable within a set of historical Latin American cultural traditions, practices, and institutions—among them the Catholic Church, the institution of slavery, racial hierarchies, Iberian corporate politics, and the juridical tradition of the Napoleonic Code. Part of the inspiration for this study is the idea of the corporate state and the body politic, based on Roman law and Catholic doctrine, which forms a commonality among Iberian cultures and their former colonies.[6] Through this corporatist ideology, which places national and collective interests above those of the individual, the political and intellectual elites of Mexico and Brazil were able to forge a centralized political model,[7] eventually developing strong top-down mythologies of national identity through *mestizaje/mestiçagem* in the twentieth century.

In his book *The Practice of Everyday Life* (1984), Michel de Certeau asserts that part of a society's history comprises myths and cultural narratives about the body, that is, an "accumulation of corporeal capital" that in turn transforms into belief systems and eventually determines the society's laws for disciplining and controlling both the social body and individual bodies: "From initiation ceremonies to tortures, every social orthodoxy makes use of instruments to give itself the form of a story and to produce the credibility attached to a discourse articulated by bodies" (149). Mexico and Brazil, like the United States, have experience with indigenous populations and with systems of slavery,[8] yet each has developed a distinct narrative about race that masks the violence of its disciplinary model.[9] For Certeau, corporeal resistance to such processes of "social inscription" is part of these cultural narratives.[10] These tensions and counternarratives drive much of the analysis here. In both Mexico and Brazil, as in other countries of Latin America, the role of race in nation building is complex, since their colonial populations were marked by an underclass that was overwhelmingly indigenous in Mexico and black in Brazil. As Juan E. De Castro points out, this racial heterogeneity meant that the European formula for nation states, which was based on a common race and language, could not be readily applied in Latin America after independence in the nineteenth century. By the twentieth century, despite the distinct political histories of these two countries, that is, the striving for liberalism in Mexico versus the oligarchical and monarchical traditions of Brazil, both had to face the disparities of their underlying racial and social realities.

Comparing Mexico and Brazil

Intellectuals in both Mexico and Brazil argue for the uniqueness of their respective cultural traditions and historical experience.[11] Mexico bases its claims of exceptionalism on its proximity to the United States, its rich pre-Columbian past and indigenous traditions, its struggles against foreign invasions in the nineteenth century, its unparalleled 1910 Revolution, and its relative political stability after 1930. Brazil often emphasizes its exceptionalism within Latin America because of its continental size, the Portuguese language, its protracted history of monarchy, slavery, and stability in the nineteenth century, and its sense of "destiny" or *grandeza* as symbolized by the modernist capital city Brasília. If Mexico focuses on the enduring legacy of the past (and a sense of historical martyrdom), Brazil projects itself as a land of unrealized potential (and messianic hope) for the future.[12] Despite the differences in the persistent codified myths of national character in these two countries—that is, Mexican "solitude" and Brazilian "cordiality"—deep structural and economic commonalities belie these mythical claims of uniqueness.[13]

In the field of humanistic studies and literature, the cultures of Mexico and Brazil are largely unknown to each other, and consequently their literary traditions and history are seldom compared.[14]

A comparative study of speculative fictions from the two most populous and influential countries of the region is especially relevant given the genre's function as a barometer of societal reactions to technological and economic change.

Mexico and Brazil share broad historical trends and patterns of economic growth and development, which influence attitudes toward technology, urbanization, and social change, all compellingly represented in the genre of speculative fiction.[15] Both nations had economies focused primarily on export from 1880 to 1930, followed by a phase known as import-substitution from 1930 to 1955, which included light industrial consumer products made for internal consumption. Ironically, late modernization did not translate into equalizing or distributive policies of wealth, even among the working classes. Douglas Graham adds that, while it is true that in the twentieth century the two countries shared high rates of economic growth, this was accompanied by steep increases in population, labor surpluses, rural poverty, income inequality, and regional income disparities (13–14). By the 1970s, the social contract between the state and the people was in crisis, as economic expansion did not signify inclusion, participation, or prosperity for a majority of the population, which remained stigmatized by race and poverty.

These late modernizing economies were characterized by a high degree of foreign investment, a focus on the production of luxury consumer goods rather than basic necessities, and the introduction of sophisticated manufacturing systems unable to absorb available labor, all of which intensified social inequalities. This social reality has not changed in the intervening years. In Brazil, after the dictatorship and the process of redemocratization that began in 1985, global competition from abroad, inflation, and anti-inflation policies all had the effect of reducing employment prospects for the growing urban population. Brazil has what has been called a "disjunctive democracy," because even though citizens are required by law to vote in elections, most of them are denied full rights of citizenship due to racial discrimination, police violence, and the spread of organized crime and political corruption.[16] Mexico has not fared much better, especially after the neoliberal policies enacted in the 1980s intensified in 1994 with the introduction of the free-trade treaty known as

NAFTA, an agreement whose devastating effect on Mexico's agricultural sector and other industries have caused millions to look for work in the United States (Edmonds-Poli and Shirk 305–9). In both countries, the pressures of violence, corruption, and narco-trafficking have eroded democratic institutions and the legitimacy of the state.

The Body in Literary History and Cultural Narratives

After Mexico and Brazil achieved political independence from their former colonizers in the 1820s, authors and intellectuals sought to create and define the identity of their newly forged nations through allegories that prefigure *mestizaje/mestiçagem*. According to Doris Sommer's seminal work *Foundational Fictions* (1991), the body and sexuality were codified in the plots of many canonical nineteenth-century Latin American novels, which she classified as "national romances." Sommer's study, which combines Foucault's ideas on reproductive sexuality with Benedict Anderson's explanation of the rise of nationalism, identifies national romances as reflections of the consolidation of national identity. A woman representing the land and indigenous peoples unites with a man representing European colonial power and culture to produce a child of mixed race who symbolizes a new beginning and national unity. Sommer notes that while the racial mixture implied in such pairings effectively cloaks the violence of the colonial past by making it part of a love story, it also establishes miscegenation as a distinct feature of national identity (39).

In many parts of Latin America, romanticism remained the predominant literary mode until 1880, at which point naturalist authors, influenced by Émile Zola's theories of applying scientific truths or methods to society,[17] began to advocate for measures designed to "cure" the national body. Although Comtean positivism was important throughout Latin America, Stephen Calogero has explained that it was most strongly felt in Mexico and Brazil. It appealed to newly forged Latin American nations mainly because it seemed to offer a straightforward explanation of their historical experience based on Comte's conception of the three stages of human civilization: the theological (combining sovereign power and religion), the metaphysical (embracing liberal ideas and revolution), and the positive (based on science and material progress). According to Comte, monarchies could offer order but no progress, and social revolutions could offer progress but no stability, while a society managed by scientists could promise order as well as social and material progress. By the 1870s, after the consolidation of the nation-states, certain sectors of the populations of Mexico and Brazil began to believe that they were poised to embark on the "positive" stage of Comtean social evolution, whereby an enlightened elite would govern the state based on the precepts of science, order, and progress (37–38), thereby overcoming both political instability and economic obstacles to modernization.[18]

The role of positivism In Mexico is most clearly seen in the regime of Porfirio Díaz, who came into power soon after the death of Benito Juárez in 1872. His government authorized intellectuals known as *los científicos* (the scientists) to implement political and educational reforms. Emilio Rabasa E. and Justo Sierra were among the influential educators who promoted positivist ideals. Unfortunately, their beliefs, which were heavily racialized, reinforced existing forms of power, and they even proposed to strip indigenous peoples of their land as part of a process that would allegedly lead to assimilation and progress (Suárez y López Guazo 86–87). Sierra believed that European blood was the solution to Mexico's social and racial problems (Suárez y López Guazo 89), a theory similar to the Brazilian push for racial whitening through immigration policies.

Regarding Brazil, Dain Borges points out that, since the 1870s, a combination of Comtean positivism, social Darwinism, and scientific racism influenced the intellectual and political class to diagnose Brazil's social malaise as a result of racial, moral, and physical degeneration (236–38). Their biopolitical solution took the form of racial whitening, which they proposed to promote through campaigns to attract European immigrants to Brazil. These measures are documented by Thomas Skidmore in his study *Black into White* (1974) and by Nancy Leys Stepan in her *"The Hour of Eugenics"* (1991), which portrays public health measures driven by eugenics and hygiene.

This "doctoring" of the national body reflects the assumption that medical science could cure social ills,[19] just as nineteenth-century positivist intellectuals and scientists believed that science and technology could undo the damage of years of colonial neglect, as shown by Rachel Haywood Ferreira in her study of utopias in *The Emergence of Latin American Science Fiction* (2011). For Haywood Ferreira, works of early science fiction are not just "pale imitations of imperialistic literary models" but rather unique adaptations of scientific discourse to the specific socio-cultural issues of the region (6). Haywood Ferreira documents participation in this scientific debate among intellectuals from Mexico, Brazil, and Argentina, analyzing their reflections on technology, evolution, mesmerism, and eugenics.

Interestingly, while Sommer's and Haywood Ferreira's studies examine roughly the same time period (1840–1920), they offer two very different portraits of Latin American fiction. While Sommer emphasizes the reproductive body in novels of national unity, Haywood Ferreira explores unnatural bodies resulting from scientific experimentation. Accordingly, two models of the body emerge: the first, biological, generative, and natural, and the second, technological, asexual, and artificial. My study builds on these models to illustrate how gender-fluid, cyborg, and living-dead or undead bodies call attention to the fractures and omissions that the discourses of national romance, positivism, and later, *mestizaje/mestiçagem* attempt to circumvent.

It is during this period of the 1920s and 1930s on that both Mexico and Brazil begin to consolidate their national cultures through the cultural discourse of *mestizaje/mestiçagem* and to formulate the national identities that we know today in popular festivals, iconic architecture, painting, films, sports, and music. All of these elements are meant to blend and champion harmony and certain unique national features, producing an ideology of collective purpose in support of a program of late modernization and state enterprise.[20] Despite the differing political orientations of their regimes in the 1930s, with the more left-leaning postrevolutionary government in Mexico and the more right-leaning government of Brazil's *Estado Novo*, both promoted a distinctive national identity in art and politics. Indeed, this may not be a coincidence. Mexican José de Vasconcelos visited Brazil on the occasion of the centenary of its independence in 1922,[21] and soon afterward was given the task of steering educational reform and national consolidation in Mexico after the Revolution. In his book *La raza cósmica* (1925; *The Cosmic Race*), he called for the forging of a *mestizo* national identity for Mexico. In a similar yet unrelated trajectory, after a sojourn in the United States, Brazilian anthropologist Gilberto Freyre emphasized the cultural contributions of Afro-descendants in the shaping of Brazilian culture in his landmark work *Casa-grande e senzala* (1933; *The Masters and the Slaves*), giving voice to Brazil's own brand of *mestiçagem*, more popularly known as racial democracy. In this way, *mestizaje/mestiçagem* in Mexico and Brazil became the basis for the social contract and the master signifier between the state and its citizens, dominating intellectual and political discourse throughout the twentieth century, effectively eliding gender/sexual difference, indigeneity, and blackness from national discourse.

The transition from the purely racial conceptualization of national identity in the form of *mestizaje/mestiçagem* to a more culturally based concept began with Fernando Ortiz's idea of transculturation, that is, the merging and/or convergence of two or more cultures in contact. Ignacio Sánchez Prado takes up this issue in his study "El mestizaje en el corazón de utopía: La raza cósmica y Aztlán y América Latina" (2009; *Mestizaje* in the heart of utopia: The cosmic race, Aztlán and Latin America), summarizing how the idea of "mestizaje" has been reworked by Spanish-American cultural theorists as "transculturation" (Fernando Ortiz and Ángel Rama), "heterogeneity" (Antonio Cornejo Polar) and "hybridity" (Nestor García Canclini). These reformulations were a way of addressing the question of national identity from a more postmodern and multi-voiced societal perspective (382). In Brazil, Alfredo César Melo uses similar terms to talk about Brazilian culture and its "disposições sincréticas e transculturativas" (289; transcultural and syncretic dispositions). His article illustrates how such terms, derived from Freyre, still persist in Brazil's cultural discourse and circulate—at times uncritically—among members of the Brazilian intelligentsia. Other critical voices include Paulo Moreira, who cites commonalities between Vasconcelos's cosmic race and Freyre's racial democracy that have led to an idealized view of the colonizer (24). Jorge de Klor de Alva asserts that Vasconcelos's idea of the cosmic race has promoted Mexico's "cultural amnesia" by denying the violence of the conquest (see "The Postcolonial Latin American Experience" 257). Similarly, Abdias do Nascimento states that racial democracy has been used to justify the monopolization of power by whites without extending the rights of citizenship to the non-white population (380).

Conceptualizing Resistance: Ethos Barroco and Biopolitics

In the 1980s, Uruguayan critic Ángel Rama was first to apply transculturation to literary and cultural studies, because it allowed theoretical reformulations of the role of traditional cultures and modernity in a new national narrative based on agency (De Castro 4–5). Rama's application of literary transculturation, according to Juan E. De Castro, "implies a sophisticated vision of political reality in which resistance to international and globalizing capital is not limited, as it frequently is in dependency theory, to the utopian possibility of revolution, but rather is to be found in the everyday actions of individuals and social groups" (7).

Rama's interest in the representation of embodied actions as narrative—particularly in terms of resistance—is similar to the sense of agency of the "ethos barroco" as propounded by Ecuadoran philosopher Bolívar Echeverría, which he derived from his own experience of Latin American culture and society. The concept of *ethos barroco*, in my view, is uniquely suited to capture and explain the cultural and literary portrayals of the body in Mexican and Brazilian speculative fiction, so I will be returning to it repeatedly throughout this study.

As a Marxian thinker, Echeverría is mainly interested in the baroque ethos that has been developed by subaltern classes in Latin America, as an attitude that enables them to survive and prosper in the face of the historic injustices and economic difficulties that have been imposed on them by colonialism and capitalism. Echeverría describes four assumptions or attitudes (Greek, *ethe*) that relate to capitalist modernity, the first two of which embrace capitalist tradition, while the second two imply resistance ("Ethos" 19–22).[22] The first, the "realist" ethos, claims that life exists for the sake of capital, subordinating all work, objects, and creativity to its paradigm of profit, while the second, the "romantic" ethos, emphasizes the creative side of entrepreneurial invention while virtually ignoring the negative aspects of capitalism. The third, or "classical" ethos, suggests resistance in that it laments capitalism's transition from the concrete correlation of the use value and market value of objects (such as the shoes that can be worn or sold) to ever more abstract forms of capital that increase economic disparities. It is the fourth, or "baroque," ethos that Echeverría identifies as key to understanding Latin American society because it contests and resists capitalism in a dialectical way that ultimately leads to social change:

> El *ethos* barroco no borra, como lo hace el realista, la contradicción propia del mundo de la vida en la modernidad capitalista, y tampoco la niega, como lo hace el romántico; la reconoce como inevitable, a la manera del clásico, pero, a diferencia de éste, se resiste a aceptarla, pretende convertir en "bueno" al "lado malo" por el que, según Hegel, avanza la historia. ("Ethos" 21)

> *The baroque ethos does not erase, as does the realist ethos, the contradiction pertaining to the world of life in capitalist modernity, and neither does it reject it, as does the romantic ethos. It recognizes it as inevitable, as does the classical ethos, but, unlike the latter, resists accepting it, endeavoring to change the "bad side" into good, through which, according to Hegel, history advances.*[23]

Echeverría characterizes the baroque ethos as a "combinación conflictiva de conservadurismo e inconformidad" (26; conflictive combination of conservatism and nonconformity) because its acceptance of the predominant realist ethos is only apparent. It contests capitalism's strict definition of profit by suggesting a method of navigating and negotiating social reality in such a way as to avoid direct conflict or confrontation with the powers that be, while pursuing social survival and economic advantage ("Ethos" 35).[24]

The baroque ethos, as a set of strategies and attitudes forged by subalterns and their allies as a way of surviving on the margins of capitalism, is both conservative and nonconformist or even rebellious.

In accordance with the baroque ethos, many social behaviors that appear to be overly complex and inefficient can be seen to be the result of the social contradictions of capitalism that streamlined or modern economies try to hide. In his essay "As ideias fora de lugar" ("Misplaced Ideas"), Roberto Schwarz shows how, even today, Brazil gives the impression of "contrastes rebarbativos, desproporções, desparates, anacronismos, contradições, conciliações" (19; "unmanageable contrasts, disproportions, nonsense, anachronisms, contradictions, compromises" [41]) that contrast with the smooth-running structures and efficiency of modern societies. Brazil's "baroque" social relations reveal its layered, disparate cultural elements, including traditional patriarchal structures, codes of modernity, and bureaucratic institutions that include popular or religious festivals and other cultural practices that defy capitalist logic and hegemony. At the same time, Echeverría warns us that the practices, beliefs, folklore, and literary trends that make up so-called magical realism should not define Latin America, because this would relegate it to the realm of premodern or insular societies, thereby dismissing or denying the reality of the baroque ethos and Latin America's central role in the history and spread of global capitalism ("Ethos" 28–29).[25]

Echeverría chooses to focus on Latin America's distinct approach to survival within capitalism and modernity. Accordingly, he rejects violent revolution as a possible alternative pathway, as he considers the idea that societies can be reconstructed from pure or new beginnings to be a myth.[26] He shows that, historically, in the face of plagues and violence during the early phase of colonization in the New World, colonizers and subalterns who were abandoned to their fate were able to survive by forging new transcultural paths and combinations, including racial ones ("Ethos" 34). In his analysis of racial *mestizaje* and the hierarchical connotations of syncretism that place European values above others, Echeverría coins the term *codigofagia* (literally, code-eating) to explain how cultural codes were actively devoured and reformulated by societies in a way that challenged the dominance of a single group or class ("Ethos" 32). The interdependence and reformulation of cultural constructs and concepts serve to break down binaries and hierarchies in a way that has much in common with

Ortiz's transculturation and also with Brazilian Oswald de Andrade's *antropofagia* (literally, man-eating), which advocates repurposing the culture and tools of the colonizer in order to produce original art and expression.[27] It is significant that both Echeverría's *codigofagia* and Andrade's *antropofagia* have embodiment in common.

I hope to show in the four chapters that make up the body of this book that Echeverría's concept of the baroque ethos is uniquely suited to serve as an overarching theory—supported and complemented by cybernetics, sexuality, and biopolitics—to explain the treatment of the body in Mexican and Brazilian speculative fiction. Below I present the main outline of its application to cyborgs, sexualities, zombies, and vampires.

Contemporary discussions of cyborgs owe much to two principal theorists who seek to imagine new ways of thinking about cybernetics and the body: N. Katherine Hayles and Donna Haraway. In *How We Became Posthuman* (1999), Hayles has argued that informational technology has brought about a "systematic devaluation of materiality and embodiment" (48), replacing it with virtual reality and its "disorienting, exhilarating effect" (27). Hayles also explores different representations of mind and spirit, noting that the mind is consistently privileged over the body in the Western tradition. In "A Cyborg Manifesto" (1991), Haraway identifies the cyborg as a feminist possibility, a new construct that sidesteps the pitfalls of gender roles and essentialist myths of embodiment. She views the cyborg as a being in a post-gendered world, "rejoicing in the illegitimate fusions of animal and machine" (176). While Haraway does not refer to baroque cyborgs per se, she does mention postcolonial societies and their need to use elements of technology and humanity in a new, recombinant way. Hence, cyborgs could make possible a new conception of the self, helping overcome the sense of Otherness caused by capitalist domination in order to express mixtures of past and present, male and female, black and white, popular and elite (161–62). The emphasis on the body and survival are in harmony with Echeverría's baroque ethos of resistance while surviving within capitalism.

Before embarking on a discussion of the body and sexuality in Mexico and Brazil, it is important to recall that both cultures operate on codified but fluid racial and sexual identities. Given patterns of conquest and colonization in both countries, racial mixture became a social reality beginning in the sixteenth century. Although no specific legislation existed barring miscegenation, racial hierarchies, prejudice, and social taboos continued to characterize both cultures. Joshua Lund observes that, in Mexico, racial mixing is combined with a series of social practices that play into racial politics: "Mestizaje is race made culture, or conversely, culture made biological: the bio-politization of race" (39). In other words, *mestizaje* makes race culturally explicit but does not overcome racism. In his summary of policies of land reform and other key historical moments, he shows that while the indigenous population is invoked as a crucial symbol, ultimately the policies are assimilationist and discriminatory (95–96). Similarly, Roberto DaMatta explains that gradations and nonbinary logic often blur the implicit racism and classism of Brazilian society, which uses a series of physical and economic characteristics as criteria to discriminate against supposed equals (*O que faz* 46–47). The reality of miscegenation was in stark contrast to the situation in the United States, where it was banned outright.

We find a similar situation with regard to homosexuality: whereas homosexuality was criminalized in the United States, England, and Germany,[28] post-independence Mexico and Brazil adopted the Napoleonic code, which generally tolerated homosexuality. As Pablo Piccato summarizes, in Mexico, some historians argue that nonheteronormative practices escaped the purview of literature and science during the nineteenth century, until it was "invented" as a crime in 1901 (90). Until then, as Victor W. Macías-González notes, homosexual men were not generally subject to criminal prosecution as long as they did not offend public sensibilities (133).[29] The same could be said of Brazil, according to João Silvério Trevisan (166–68) and James N. Green (*Beyond Carnival* 22–23). Green notes that homosexual social spaces and practices in late-nineteenth-century Rio de Janeiro were documented in medical literature but were neither investigated by the medical community nor criminalized by the justice system ("Doctoring" 193–95).

Eventually, however, homosexuality was identified as a problem in these societies. As Michel Foucault outlines in his *The History of*

Sexuality (1980), although no specific laws prohibited homosexuality, the nineteenth-century classification and "pathology" of sexual practices evolved into a pretext for controlling nonnormative bodies and nonheteronormative sexualities. In the case of Mexico, Piccato notes a preoccupation surrounding homosexuality and lesbianism in male and female prisons in Mexico City, when a series of works dating from 1904 to 1910 were published by journalist Carlos Romagnac featuring interviews with inmates (88). Picatto notes that homosexuality obsessed authorities who attempted to use the transgressions of gender as evidence of the threat of the erotic practices of the urban poor to proper Porfirian society (101).

These kinds of anxieties coincide with fears of racial degeneration and the resulting desire to construct a national identity based on the strength of racial mixture and *mestizaje/mestiçagem*. In their introduction to *Gender, Sexuality, and Power in Latin America since Independence* (2007), William E. French and Katherine Elaine Bliss observe that, for nearly two hundred years, authorities in Latin America have correlated specific sexual and racial constructions with nation-building and identity (23). They also note that the work of Judith Butler and queer theory have served to deconstruct essentialist categories such as heteronormativity and have "uncovered more diversity in categories such as homosexual and lesbian," which in themselves may function as "social fictions, covering over or hiding a diverse array of embodiments and sexual desires" (19). These surmises are confirmed by the diversity of embodiments I have found in Mexican and Brazilian speculative fiction, which feature fluid categorizations of gender, desire, and even species.

The concept of the baroque ethos forms an overarching approach to nonnormative sexualities, in the sense that those who flouted societal norms were forced to find ways around the increasingly punitive attitudes toward their behavior. Other theorists have contributed to this issue. João Nemi Neto, for example, argues that Latin American homosexuals have had success in negotiating their sexuality in a way that is less confrontational than the American concept of "coming out." Borrowing from Nemi Neto, I use the concept of the "baroque queer" as a way of explaining resistance through sexuality and gender in speculative fiction.

As for resistance expressed through the living dead and the undead, that is, zombies and vampires, I am interested in overlap between Echeverría's baroque ethos and the biopolitical theories developed by Roberto Esposito. Esposito's ideas of biopolitics, based on concepts of community and immunity, are helpful in understanding resistance in the case of bodily invasions. In his 2002 *Immunitas* (trans. from the Italian in 2011 as *Immunitas: The Protection and Negation of Life*; citations refer to 2011 edition), Esposito notes that individuals in a given body politic are joined by a common *munus*, that is, duties that they owe to each other and that bind them into communities (5). These duties, however, require that individuals give up something, namely freedom and individual rights or identity, in order to belong to the community and enjoy its protections. Esposito then sketches out the implications of this obligation. First, it is ironic but true that the community protects its members from violence through the promise of violence, indeed by claiming a monopoly on violence (9–10, 29), for example, by authorizing the police to imprison or even kill citizens who ignore community standards of behavior as formulated in laws (21). Second, it is inevitable that certain members of the community will try to enjoy its protections without making the sacrifice implied by the *munus*, they will seek immunity from laws, thereby gaining privilege (6). We can imagine that elites might use powerful connections for this purpose, leading to gross inequality at a minimum and outright corruption in some cases, while other elements of society might simply ignore the obligations, essentially engaging in evasive activity. Ideally, the community would protect itself against both of these strategies.

A third implication is that, precisely as our immune systems protect us from pathogens and viruses that have penetrated into the body, the body politic must protect its members from outside threats that make their way into the population (Esposito 7). Our biological immune systems must react to pathogens such as parasites, microbes, and viruses, while the body politic's immunological reaction is triggered by people whose ideas and actions are considered, for various reasons, to be dangerous to the whole.

Finally, Esposito takes the metaphor one step further by noting that reactions by security forces to perceived threats may escalate to the point that they themselves become a threat to the communities that they are intended to protect, in a kind of "autoimmune" reaction (17). Esposito describes such overreactions as "a backward wave that in an attempt to immunize the community from the process of generalized immunization spreads death and unleashes self-destruction in a desperate (auto-immune) gesture that attacks the blood of its own body" (53).[30] Thus, for Esposito, autoimmune diseases are for the body what corrupt states, civil wars, or violent police states are for society (164).

At this point I would argue for extending the metaphor even one step further, adopting the perspective of the parts of the body that are targeted by autoimmune reactions. When certain groups within the body politic find themselves under attack by a police apparatus, they may engage in whatever counter-immune strategies they deem necessary in order to defend themselves. This is clearly seen in some Mexican and Brazilian speculative fiction, in which targeted populations, especially the poor and marginalized, are portrayed as zombies. In other words, zombies can represent both the invading pathogen from which the community must protect its individual members and the targeted individual members trying to protect themselves from an autoimmune overreaction by the community.

Esposito appeals for a different kind of immune response, one based on a conception of the body (and the body politic) as a "functioning construct that is open to exchange with its surrounding environment" (17), rather than bent on violent suppression of the Other, just as the body of a mother "tolerates" a fetus (169–70). This is perhaps a reaction to Giorgio Agamben's more pessimistic analysis as presented in his *Homo Sacer: Sovereign Power and Bare Life* (1998), which sets forth the basic outline of biopolitics that grew out of Foucault's ideas about the institutionalization of power by the modern state and the control and regulation of populations in society. Foucault proposed the idea that authorities ultimately deploy biopolitics by deciding whom they will let live and whom they will let die (*Society* 247), and Agamben believes that state control of bodies has progressed to the point that human beings have very little control over their lives—that is, that they are "bare life" (Greek, *zoê*) rather than full citizens (Greek, *bios*) (4). A key concept in Agamben's book is that of the *homo sacer* as defined in imperial Roman law, as a person who has been deprived of rights and can be killed by anyone without repercussions (8). He considers that modern democracy is the struggle of the disenfranchised—of the political *zoê*—to find a way back to *bios* (9). Agamben believes, however, that there is little the individual subject can do in the face of sovereign power and its thanatopolitics or "politics of death," that is, that *bios* has all but disappeared in modern society, along with the oppositions that form the basis of modern politics such as right / left, authoritarianism / democracy, public / private. Esposito, in contrast, offers a prescription for a more benevolent form of "immune response"—based on tolerance and dialogue with the Other—as an alternative to Agamben's negative vision of biopolitics.[31]

To a certain extent, Mexican and Brazilian elites have privileges that provide them immunity from the law, but the same does not hold true for the majority of the population. This is the hidden message behind *mestizaje/mestiçagem* discourse, whose claim that mixed-race people are "included" within the nation is belied by the fact that they are excluded from political privilege and power, a circumstance that Esposito calls "exclusion by inclusion" (8). Esposito's concept of immunity is of special importance for understanding resistance in texts about the living dead in Mexico and Brazil, because as disenfranchised groups targeted by the state's immune response, they may take counter-immunological measures in form of resistance to state violence and coercion. In other words, as David Dalton has suggested, subalterns may resist and even subvert *immunitas*, undermining state power and creating their own form of resistance ("Antropofagia" 5).

The Traditions of Speculative Fiction in Mexico and Brazil

Mexico and Brazil share similar trends in the history of speculative fiction, lending themselves to comparison.[32] In both countries, several key works appeared in the nineteenth century, as documented in Rachel Haywood Ferreira's *The Emergence of Science Fiction*. Utopian fiction was especially important during the period of intense urbanization at the turn of the twentieth century.[33] The rapid industrialization of the 1930s and '40s fomented the publication of landmark science fiction by Diego Cañedo in Mexico and Jeronymo Monteiro in Brazil (Molina Gavilán et al. 373, 381). In the 1950s and '60s, mainstream Mexican authors such as Juan José Arreola and Carlos Fuentes, among others, penned stories that could be categorized as science fiction, while Carlos Olvera's *Mejicanos en el espacio* (1968) inaugurated a trend of using the genre as social critique. These same trends appeared in Brazil during the 1970s, as mainstream authors adopted the science fiction genre as a way of escaping censorship by the military regime (Ginway, *Brazilian* 89–136). In the 1980s, both countries experienced a boom in science fiction, although in different ways. In Mexico, support was formal, in that the National Council for Science and Technology (CONACYT) began to sponsor science fiction prizes in Puebla in 1984, launching a generation of SF writers born in the 1960s (Fernández Delgado, "Introducción" 11–14). In Brazil, it was driven by a strong fan base, which led to the establishment of new publishing outlets (Ginway, *Brazilian* 25–27). National versions of cyberpunk also emerged in both countries in the 1990s, with a new wave of science fiction, accompanied for the first time by academic studies, beginning in the 2000s.[34]

Portrayals of bodily distortion first appear in Mexican and Brazilian speculative fiction in the late nineteenth century. This is somewhat earlier than might be expected for late-modernizing countries, and shows that at this early date these authors were already participating in the discourse of science and science fiction. I would argue that by focusing on the body—changes in sexual identity, the presence of prosthetics or artificial bodies, and the traits of the living dead and the undead—the authors of these stories were able to convey in an effective and powerful way prevailing anxieties about society, politics, and technology. Thus, in Mexico and Brazil, the transformed, artificial, or distorted body is not simply a rewriting of colonial chronicles or a product of magical realism, but rather a symptom of crisis and change in these two large, industrialized, and ethnically diverse countries.

If the Latin American tradition has not always been recognized as part of the global history of speculative fiction, it is solely because of the emphasis traditionally placed on a core of texts from the global

North, according to Australian critic Andrew Milner, who examines Anglo-American, Asian, European, and Russian examples in his study *Locating Science Fiction* (2012). Conceptualizing science fiction as a "world genre," Milner argues that Latin American science fiction sometimes falls outside the parameters of the genre's selective history or tradition (177) because it often includes a combination of overlapping elements from fantasy, utopia, science fiction, and—notably—gothic horror, a genre whose notorious fascination with the body makes it an indispensable part of any study of the body in speculative fiction.

The focus on the body is a further justification for the inclusion of gothic horror in the wider genre of speculative fiction, especially in Latin America. Persephone Braham's study *Amazons to Zombies: Monsters in Latin America* (2015), for example, encompasses monsters from the first chronicles of discovery, through canonical literary texts to contemporary film. These same themes and genres are examined in two additional interpretive studies that tend to confirm my own findings. In *The Tropical Gothic in Literature and Culture* (2016), editors Justin Edwards and Sandra Guardini Vasconcelos interpret texts from the American South, Haiti, Mexico, and Brazil as adaptations and subversions of established tropes (2), while the essays that comprise *Latin American Gothic* (2018), edited by Sandra Casanova-Vizcaíno and Inés Ordiz, reinforce the subgenre's status as a component of speculative fiction. As the editors state, the texts analyzed are "rooted in local realities" and deploy "appropriation and / or parody" as strategies of social critique (7).[35]

This conceptualization of speculative fiction contrasts with Darko Suvin's well-known 1979 definition of science fiction, which is based on the presence of a "novum" (or innovation) that provokes a reaction of "cognitive estrangement" (defamiliarization) from our known reality (4), thus eliciting a rational examination of social relations, power, and ideology. To bridge the gap to the wider genre of speculative fiction, Fred Botting suggests replacing "novum" with "monstrum." Both concepts trigger defamiliarization by blurring the usual boundaries between the natural and the unnatural, reality, and fantasy, but *monstrum* is more effective at capturing the powerful yet at times threatening nature of altered bodies that signal resistance through the baroque ethos and *codigofagia* (112). Needless to say, neither Botting nor I wish to imply that gender fluid or nonbinary bodies of actual humans are monstrous in any way. The term represents the point of view of the elites, who interpret any deviation from accepted norms as a threat to their hegemonic control of society.

Chapter Trajectory

The unique interpretation of the trope of bodily transformation in Mexican and Brazilian speculative fiction is a useful way of showing its place in the larger global genre. I take my historical cues from Haywood Ferreira's *The Emergence of Latin American Science Fiction*, mentioned above, as well as from Roberto de Sousa Causo's *Ficção científica, fantasia e horror no Brasil 1875 a 1950* (2003; Science fiction, fantasy and horror in Brazil 1875–1950), a historical survey of speculative fiction in Brazil, and Ross Larson's insightful work on short fiction, *Fantasy and Imagination in the Mexican Narrative* (1977). I glean insights on more contemporary works from J. Andrew Brown's *Cyborgs in Latin America* (2010), an examination of the cyborg body in Southern Cone literature of the post-dictatorship as well as in Mexico and the Andes, and Miguel López-Lozano's *Utopian Dreams, Apocalyptic Nightmares* (2008), an analysis of Mexico's socio-political crisis surrounding NAFTA and neoliberal policies (1992–1996). While Mabel Moraña's *El monstruo como máquina de guerra* (2017; The monster as war engine) stresses cultural icons and literary theory in defining monsters, and Braham's *From Amazons to Zombies* (2015) examines canonical literature and contemporary film from Spanish America, I tend to follow David Dalton's paradigm of *Mestizo Modernity* (2018), because he emphasizes the transformation of the human body by technology in visual, cinematic, and literary forms, using biopolitics and racial paradigms in order to critique the official discourse of *mestizaje*. In general, I have chosen a comparative approach, analyzing works of short fiction and a few select novels, both canonical and noncanonical, spanning the late nineteenth to the twenty-first century in order to give a comparative overview of the centrality of the body in Mexico and Brazil, highlighting the similarities and differences between them.

Chapter 1, "Gendered Cyborgs: Mechanical, Industrial, and Digital" examines the cyborg body principally as an allegory of late modernization in Mexico and Brazil through Echeverría's baroque ethos, supplemented by concepts of commodity fetishism and cybernetic paradigms. Structured chronologically, the chapter focuses on three key moments when gendered cyborgs are most in evidence, adding a level of estrangement to the analysis of modernity: the late nineteenth century, the mid-twentieth century and the turn of the twenty-first century. In the first section, "Female Fetish Cyborgs in the Long Nineteenth Century," I show how doll-like cyborgs portray a longed-for yet elusive modernity as described by Ericka Beckman in her 2013 *Capital Fictions*. In narratives by Joaquim Maria Machado de Assis, João do Rio, and Pedro Castera, gendered cyborgs simultaneously exemplify and undermine elite modernization by subverting or resisting dreams of wealth, civilization, and science. In the second section, "Resistance and Rebellion of Mid-Century Industrial Female Cyborgs," the artificial female body is portrayed as a series of feedback loops, as described by N. Katherine Hayles in *How We Became Posthuman* (1999). In tales of consumerism and efficiency paradigms by José Arreola, Dinah Silveira de Queiroz, and Alfredo Cardona Pena, gender and sexuality become a means of expressing the baroque ethos, in that cyborgs are shown disrupting, destabilizing, or resisting conventional societal expectations, economic growth, and consumer culture. In the 1970s, Caio Fernando Abreu and Emiliano González create cyborgs that express intelligence and emotional states characteristic of Hayles's theories of "reflexivity," in which artificial systems incorporate and express human consciousness. These cyborgs, armed with self-awareness, actively resist the commodification of sexuality and the Foucauldian apparatuses of state power. The final section, "Neoliberal Cyborgs and Gender," covers the post-1980s period, in which digital information, surveillance, and bodily control increasingly destabilize national boundaries and the division between human and artificial life forms. Cyborgs evolve into posthumans who question and resist their roles as compliant bodies in the outsourcing of labor, and also resist divisions between human and artificial, male and female. In works by Pepe Rojo, Bef (Bernardo

Fernández), Roberto de Sousa Causo, and João Paulo Cuenca, cyborgs disrupt and defy surveillance and digital data used by corporate or criminal networks to control them. Thus the gendered cyborg body remains as a sign of the resistance of the baroque ethos against neoliberal models of digital interfaces and modernity.

Chapter 2, "The Baroque Ethos, *Antropofagia*, and Queer Sexualities" is organized thematically and combines Echeverría's baroque ethos with João Nemi Neto's concept of the "anthropophagic queer," which itself combines concepts from American studies on sexuality and queerness with Silviano Santiago's nonconfrontational tactics or "wiliness" as a way to affirm a queer identity. Following Judith Halberstam's 2011 *The Queer Art of Failure*, in which she examines gender fluidity in Pixar movies to critique neoliberal individualism and heroic narratives of achievement, I offer alternatives to the heteronormativity or romantic pairings so celebrated in mainstream culture. The section "Women Warriors" includes works by Machado de Assis, Gastão Cruls, and Carmen Boullosa, who examine gender performativity, war, violence, and colonialism. As women warriors, the protagonists of these texts come closest to traditional heroic masculinity, only to find themselves in a queered "between-space," where they either negate or question heroic discourse. The second section, "Women in Men's Bodies," examines texts by Amado Nervo, Coelho Neto, and Roberto de Sousa Causo, in which the transmigration of souls and other occult practices place women's minds into the bodies of men, invading homosocial spaces with destabilizing or even fatal consequences. These queered protagonists choose either to leave their hosts or to join alternative communities in an evasive strategy of survival typical of the baroque ethos. In the final section, "Women as Other Species," I examine works by Efrén Rebolledo, Diana Tarzona, and Aline Valek in which bodily transformation and women's identification with insects, reptiles, amphibians, and other aquatic forms of life serves as a means of representing gender / queer communities outside of heteronormative conventions. Such queering and identification with cold-blooded or invertebrate species avoids facile anthropomorphic interpretations and can be best understood as acts of resistance typical of a "baroque queer."

Chapter 3, "Trauma Zombies, Consumer Zombies, and Political Zombies" examines three types of zombies. Since zombies are usually portrayed as mindless, lifeless, and lurching, that is, the "living dead," they can also be understood as harbingers of change, embodying the viral, uncontrolled paradigms of economic or political upheaval. After the first section, "An Overview of Zombie Criticism and Typology," I examine stories about humans who act like zombies after experiencing economic devastation. These "trauma zombies" include the fictional inhabitants of Brazil's "dead cities" after the end of the coffee boom at the turn of the twentieth century, who, as proto-zombies, easily succumb to gold fever or the virus of untold wealth in stories by Lima Barreto and Monteiro Lobato. Iterations of the human or trauma zombie return at the end of the twentieth century and beginning of the twenty-first, after neoliberal policies have taken their toll on both Mexico and Brazil, as seen in works by Diego Velázquez Betancourt, André de Leones, and Homero Aridjis. In "Consumer Zombies: Baroque Resistance in Parody and Play," we find humor, profanation, and satirical critiques of consumer society in tales from the 1940s and '50s by Ortiz de Montellano, Manuel Becerra Acosta, and Carlos Drummond de Andrade. I argue that from the mid-century on both Mexico and Brazil seek to "immunize" themselves from the vulnerability of global markets, building their state industries along with apparatuses to protect themselves from internal dissent, another immunization paradigm. Finally, in "Political Zombies and Esposito's *Immunitas*," I examine tales of zombie resistance by Bernardo Esquinca, Erico Veríssimo, and Lygia Fagundes Telles, who write about segments of the population that "zombify" themselves as a form of counter-immunity against oppression, resisting or toppling authoritarian governments in the 1960s and '70s. A final, more contemporary tale of resistance can be found in the work of Karen Chacek, whose somnambulist zombies immunize against the viral effects of economic consumer society and the enforcement of economic productivity.

Chapter 4, "Vampires: Immunity and Resistance" is dedicated to monsters whose fully functional intelligence and personalities—contrasting with zombies, which are essentially animated corpses—is best captured by the term "undead." In the first section, "History of the Vampire Figure," I note that approaches to the vampire in Anglo-American culture may be applied to vampires in the New World as well, focusing either on gender and sexuality or on capitalism and colonialism. In "Vampires in Domestic Space: Immunological Resistance," I examine stories by Alejandro Cuevas and Carlos Fuentes, whose foreign vampires embody Esposito's immunological threat to the health of the national body politic imported by outsiders or foreigners. The next section, "The Vampire, Female Defiance, and the Tropical Gothic" examines the role of female characters who face vampiric forces in the structures of the traditional home or ancestral plantation and in the Catholic Church in works by Amparo Dávila, Lúcio Cardoso, and Gabriela Palafox. I end the chapter with *"Bíos and Citizenship: The Vampire and the Social Other,"* in which vampire characters uncharacteristically begin to wreak havoc on former colonial masters and side with the oppressed by acting on behalf of the marginalized. In tales of colonial revenge by Gerson Lodi-Ribeiro, Gabriel Trujillo Muñoz, and André Vianco, vampires free the oppressed from the ills of the past, present, or future. Other iterations include human/vampire mediators, who, in novels by José Luis Zárate and Giulia Moon, present an alternative biopolitical model based on alliances forged among diverse members of society rather than on overt resistance to state power.

CHAPTER I

Gendered Cyborgs
Mechanical, Industrial, and Digital

This chapter traces the technological augmentation, enhancement, and even replacement of the human body in Mexican and Brazilian speculative fiction. In order to simplify the presentation, I employ the word *cyborg* to denote several types of beings, including altered humans, whose bodies have been implanted with prostheses; cyborgs proper, which combine cybernetic and organic systems to an extreme degree; and robots or androids, which are fully mechanical or artificial beings. Examining the texts from Mexico and Brazil over the past hundred years that feature these kinds of beings, we will see how technologically altered bodies offer insight into personal boundaries and the concept of embodiment, especially as these concepts relate to changing social and political institutions and national identity.

I will examine cyborg texts from three key economic periods.[1] The first covers the primarily export-based economy that extends from 1870 to 1910 and features female proto-cyborgs, that is, automatons that function as fetishes of modernity. The second, which focuses on postwar industrialization from 1945 to 1980, introduces female robots and cyborgs that are industrial products intended for internal consumption and markets. The third period encompasses the neo-liberal period, the mid-1980s to the present, focusing on cyborgs and posthumans as compliant bodies and outsourced labor in a globalized world. All of these entities—doll-like automatons, industrial robots, and posthuman cyborgs—capture the dilemmas of gendered labor and social inequities while simultaneously serving as a voice for those who have been silenced in the national collective memory. I believe that the female or ambiguously gendered cyborg offers the most original and resistant iteration of this figure, estranging the body in mechani-cal, industrial, and digital paradigms of work and modernity, and for this reason I will focus on this type of cyborg throughout the study.

Thematically, the cyborg body may be used as a vehicle for explor-ing the borders between masculine and feminine, organic and artificial, and traditional and modern. Notably, the cyborgs and other altered human bodies appearing in Mexican and Brazilian speculative fiction are often arbitrary assemblages of disparate parts, with low-tech pros-theses and inefficient interfaces that are repurposed for improvised functions. This feature, which reflects the unevenness of technolog-ical transfers, can be seen as subverting paradigms of productivity and efficiency. Gendered cyborgs in both Mexico and Brazil are espe-cially effective at capturing the social, racial, and technological con-tradictions of late development and modernization, adding an extra layer of estrangement to social relations. For this reason, although they can be analyzed in accordance with N. Katherine Hayles's con-cepts of three stages of cybernetics,[2] they also provide an excellent illustration of Ecuadoran philosopher Bolívar Echeverría's "baroque ethos," because they resist the methods and products of capitalism while participating in it.

As I explained more fully in the Introduction, Echeverría believes that the subaltern classes of Latin America have developed an "ethos" that can make their lives bearable, even satisfying, in the face of histo-ric injustice and economic difficulty. He defines an ethos as a paradoxi-cal concept that can mean either "shelter," in the sense of tradition or belonging, or "resistance," in the sense of challenge, for example to a given restriction assigned by society ("Ethos" 26). As a cultural strategy of the subaltern, the baroque ethos exemplifies an attitude that "acepta las leyes de la circulación mercantil . . . pero lo hace al mismo tiempo que se inconforma con ellas" ("Ethos" 26–27; "accepts the laws of mercantile circulation . . . while simultaneously refusing to conform to them" [Gandler 301]).[3] In general, the baroque ethos subverts capitalism's strict notions of profit, developing attitudes of conservatism and nonconformity in dealings with social reality, in which subalterns appear to conform yet resist or subvert the dictates of the more powerful (Gandler 301–2).[4]

Gazi Islam has examined the continued relevance of the concept of cultural *antropofagia*, which resembles Echeverría's *codigofagia* of cultural absorption. In Brazilian society, Islam notes that such knowledge is embodied and encoded by individuals and social institutions in a way that rejects the separation between mind and body that has been central to modern thought since the Enlightenment (172). Islam illustrates this by citing the example of women's police stations in Brazil that refuse to publish mission statements because they recognize that theirs are hybrid or in-between institutions based on patriarchal structures but "manned" by women whose feminist principles move them to protect other women from violence. *Codigofagia* is useful in analyzing female cyborgs who may appear to conform to gender expectations but whose minds resist subordination by others. These characters also illustrate the importance of gender in analyzing structures of modernity in a patriarchal culture, where such adaptations do not fit into conventional feminisms, but rather are highly adapted to a social context and the power structures of race and class.

Islam also considers the triadic structure of the "language of difference" to be a reappropriation, in an active sense, of resistance, similar to that propounded by the baroque ethos and *codigofagia*:

> The corporalized nature of anthropophagy, finally, contrasts with the identity politics of some postmodern discourse. . . . Different from political correctness, which sees language as a form of domination, anthropophagy recognizes language as a tool of desire, opens up a hybrid liminal or "third space" (Bhabha, 1994) where the dyad "colonizer-colonized" can be unsettled though changing the meanings of colonial language. (173)

What I wish to argue here is that the cyborg body and the baroque ethos articulate this third space, because they conflate human and machine, male and female, colonizer and colonized, traditional and modern. Gendered cyborgs tell stories that allow us to understand the how and why of apparently contradictory behaviors, focusing on people who do not conform to standards of modernity because they do not benefit directly from them, but rather choose to adapt them to their needs in order to survive.

By analyzing cyborgs from three distinct periods, I offer a historical sampling that emphasizes gender and embodied labor, using Echeverría's concepts to highlight the distinct baroque character of Latin American speculative fiction. As a way of understanding late nineteenth and early twentieth-century cyborgs, I begin by analyzing proto-cyborgs and the baroque ethos in light of Ericka Beckman's *Capital Fictions* (2013), a study of the Latin American literature and culture of the export age (1880 to 1920). Beckman bases her analysis on Marx's concept of commodity "fetishism," by which a product's market value gives it an aura of mystical desirability, to illustrate how Latin American elites were seduced by promises of export-based financing for modernization. Beckman shows the repeated failure of export economies to realize the economic elites' dreams of modernity in Guatemala, Colombia, Argentina, and Cuba in the late nineteenth and early twentieth centuries. While Beckman emphasizes the predominant realist ethos and the strength of external markets for elite investors, I focus on the baroque ethos during the same time period, examining internal markets and using cyborgs as allegories of labor. I take my cue from Beckman in also highlighting what "capitalist fictions" try to hide: the complexity of systems of production described in these nineteenth and early twentieth-century techno-fictions.

In the mid-twentieth century, the economies of postwar Latin America provide an interesting scenario for examining the industrial cyborg and the baroque ethos, as they begin to develop their own internal markets through state-sponsored import substitution. N. Katherine Hayles's cybernetic paradigms, as outlined in *How We Became Posthuman* (1999), are of use here. She explores different representations of mind and spirit, noting that the mind is consistently privileged over the body in the Western tradition. In her analysis of information science and its portrayal in science fiction texts, Hayles distinguishes three phases in cybernetic theory. The first is the homeo-static cybernetic model of combined living and mechanical systems described in Norbert Wiener's 1949 book *The Human Use of Human Beings*. Based on studies of homeostasis (the tendency toward stable equilibrium among interdependent elements), it relies on feedback loops for self-regulation. The second phase of cybernetics, dated during the 1960s and '70s, is characterized by "reflexivity," according to which beings "create" their own reality from inside the cybernetic feedback loop based on the internal adjustments within the system, which recognizes humans as part of the system. In the third phase of cybernetic theory, Hayles claims that pattern recognition and random-ness produce what she calls "flickering signifiers" of data, which privilege cybernetic, disembodied knowledge and cyberspace over face-to-face contact and other humanistic forms of knowledge (35).

Industrial cyborgs tend to fall within the first two paradigms of cybernetic homeostasis and reflexivity. Armed with these tools, I analyze how mid-century gendered cyborgs in Mexico and Brazil work within the informational and industrial feedback loops, both as products of consumption and as agents of resistance. I note that, as domestic space becomes a metaphorical factory for production and consumption, gendered cyborgs initiate sexual couplings that pro-mote, but most often interfere with, the smooth functioning of the capitalist feedback loop, disrupting normative concepts of modernity.

No study of gendered cyborgs would be complete without a consideration of Donna Haraway's "A Cyborg Manifesto" (1991), which conceptualizes the cyborg as a feminist possibility, a self-created, self-engendered female, free from the demands of capitalism and the Freudian neuroses caused by traditional family life (150). We find, how-ever, that despite their rebellious nature, neither Brazilian nor Mexi-can female cyborgs fulfill this feminist vision for independence, as they all yearn for traditional ties and emotional acceptance through romantic partnerships or family structures, as part of the baroque ethos.[5] These gendered cyborgs also have much in common with what Hayles characterizes as "reflexivity," as described above.

Several previous studies have explored the gendered cyborg in Mexico. In his 2000 *Science, Technology and Latin American Narrative*, Jerry Hoeg identifies Carmen Boullosa's novel *Duerme* (1994) as the inspiration for his term "cybermestizaje," characterizing the work's protagonist as a "hybrid fusion" (99). Boullosa further develops the female cyborg in her novel *Cielos en la tierra* (1997), but, as J. Andrew Brown notes, in this story the cyborg yearns to recapture a human identity and escape the dehumanizing effects of technology (54). Thus, *Cielos en la tierra* negates Haraway's liberating idea of a posthuman feminism, since by the end of the novel, the cyborg rejects her society as a technological nightmare, in which "posthuman bodies merely reconstitute an oppressive social order rather than subvert it" (Brown 56).

As we shall see, these gendered cyborgs are not monsters of capitalism but rather human/machine hybrids that allow us to see the new iterations of the human in the posthuman that resist and occasionally disrupt the system, at times conveying an implicit utopian possibility of social redefinition and renewal.

Female Fetish Cyborgs in the Long Nineteenth Century

I begin with a story by Brazilian master Machado de Assis, whose work has been insightfully analyzed by Roberto Schwarz, who first set forth his concept of "ideias fora do lugar" (misplaced ideas) in his study on Machado, *Ao vencedor as batatas* (1977; To the winner go the potatoes). He notes that, in their efforts to match European civilization and culture, nineteenth-century Brazilian elites adopted liberal ideas that ignored the reality of their slave-based coffee economy in order to embrace ideals of industrialized economies and free labor (14–15), principles that actually formed the political basis for Brazil's 1824 Constitution. Many of Machado's characters skillfully navigate this contradiction, while others exhibit a metaphorical blindness and deny slave-based reality altogether, and a few, such as Rubião in *Quincas Borba* (1890), eventually succumb to insanity and death.

Students of Brazilian science fiction will be interested to know that Machado penned one of Brazil's first proto-cyborg stories in "O capitão Mendonça" (1870; Captain Mendonça), although it eventually turns out, as in many of Machado's stories of the fantastic, that the events were part of a dream. Loosely based on E. T. A. Hoffmann's 1816 story "The Sandman,"[6] Machado's "O capitão Mendonça" is the story of a young man named Amaral who becomes obsessed with Captain Mendonça, an inventor, and his invention, a female automaton. The story is framed by Amaral's evening at a theatrical performance, after which he decides to accept an invitation to dine at Mendonça's home. The house, with its dark rooms, taxidermied animals, and labs for alchemy experiments and other sinister operations, is reminiscent of a gothic horror setting. Amaral describes the house as having a "purgatorial" character (187), thus casting Mendonça as a Faustian figure whose far-away look, bushy eyebrows, and eccentric manner recall that of a mad scientist. Soon Amaral meets Augusta, a beautiful green-eyed female automaton whom Mendonça introduces as his daughter. When Amaral notes that her green eyes are the same color of those of a stuffed owl, Capitão Mendonça asks if Amaral would like to examine them, removing the eyes of his "daughter" for Amaral to inspect. Her face then appears skull-like to Amaral, who remarks that she looks like "uma caveira viva, falando, sorrindo, fitando em mim os dois buracos vazios" (188; a living skull, talking, smiling, fixing her two empty sockets on me). He also comments on how the eyes are still looking at him in Mendonça's hands, as if in a conscious way: "Daquele modo, as duas mãos do velho olhavam para mim como se foram um rosto" (188; That way, the two hands of the older man looked at me as if they were a face), suggesting a strange seduction and manipulation. Yet despite these uncanny, deathly images, Amaral returns repeatedly to the house, and after Mendonça gives him a diamond, Amaral proposes to take Augusta as his wife. Mendonça approves the union under the proviso that Amaral undergo a brain operation to match his intelligence to that of his future wife. As he is immobilized by Augusta for the procedure and Mendonça prepares to perforate his scalp, Amaral suddenly awakens to find that he has fallen asleep in the Teatro São Pedro, where he realizes that he has dreamed the entire episode.

As I have discussed elsewhere, the recurrent colors of gold and green of Brazil's royal coat of arms in the story associate the character Capitão Mendonça with Dom Pedro II in the role of the emperor's shadow side (Ginway, "Machado's Tales" 214–15). The captain's recent arrival from Rio Grande do Sul would associate him with both the War of Oribe and Rosas (1851–1852) and the Paraguayan War (1865–1870), reminding readers of the importance of the wars fought in the River Plate area in establishing Brazil's hegemony in the region, consolidating the reign of Dom Pedro II and the idea of Brazil as nation. Portrayed as a peace-loving monarch whose erudition and love of science were well known, his associations with slavery, war, and violence have generally been suppressed in Brazil's collective memory. Through the figure of Capitão Mendonça, Machado de Assis may be suggesting that the reign of Pedro II may not be as peaceful and stable as it appears. Readers have hints of this in the sinister image of Augusta whose head becomes a talking eyeless skull, an image of death—perhaps of repressed historical memory—which Mendonça himself quickly displaces with the marvels of science and a product of commodity extractivism, namely, diamonds.

In order to induce Amaral to marry his daughter, Mendonça initiates a scientific procedure that produces a diamond from coal as if by magic. Notably, the diamond is a perfect example of a commodity fetish, since it has no use value but high market value, and its sparkling perfection distracts from the labor behind its fabrication. The gift of the diamond persuades Amaral to marry Mendonça's mechanical daughter, dazzling him with the promise of wealth and the marvels of "science." Significantly, the words *diamond* and *coal* appear over twelve times each in the story, reinforcing an obsession with mineral commodities that formed the slave-based export economy of Brazil during the eighteenth century in the mining state of Minas Gerais, where a first independence movement, A Inconfidência Mineira, took place in 1789. As a rebellion begun by intellectuals outraged by the colonial system of the Portuguese Crown, the revolt recalls the underlying ideals of liberal democracy that were later undermined once Brazil gained independence in 1822; both Dom Pedro I and his son Dom Pedro II retained an absolutist hold on power, a theme hinted at by Mendonça's persuasive manner and need to impose his will.

In Machado's story, Amaral must resist two forms of temptation or commodity fetishes: first, diamonds, an extractive product of slave labor, and second, Augusta, a product of modernity and science.[7] However, in the end Amaral is able to draw on the baroque ethos and snap himself out of what Beckman would call an "export reverie" and awaken to resist temptation. As a proto-cyborg associated with diamonds and commodity fetishism, Augusta's cyborg body reveals two sides of Brazilian society; her skull-like head functions as a reminder of death and those bodies sacrificed to war and slave labor, while her graciousness sparkles like a diamond, recalling the "civilized" manners of the upper classes who use polite subterfuge to disguise their iron will backed by the power wielded by the monarchy. The fact that Augusta merrily continues to speak when her eyes are removed is suggestive of a "living death," while her desire to alter her fiancé's brain reinforces ideas of slavery and zombification. Augusta, as a proto-cyborg, is a fascinating intersection of gender, labor, and commodity extraction, providing an example of *codigofagia* that is at once archaic (demanding tribute) and futuristic (promising modernity). As a baroque cyborg, Augusta incorporates all social strata and the contradictions of Brazil's nineteenth-century society during the height of the economic boom, which is both glimpsed and resisted by Amaral. Machado continued developing this dizzying obsession of export reverie and insanity in his later novels.[8]

Some forty years later, similar imagery associating a female proto-cyborg and death appears in João do Rio's 1910 story "O bebê de tarlatana rosa" (The baby in pink tulle), a work that Alexander Meireles da Silva characterizes as belonging to the subgenre of "gothic science fiction" in Brazil (16–17).[9] The gendered cyborg of this tale represents a different type of baroque ethos, since although enchanting, she is a less sophisticated cyborg of the lower classes. The title character, a young woman referred to only as the *bebê*, (the baby or darling) wears a prosthesis in the shape of a nose that gives her cyborg-like qualities. When the upper-class narrator, Heitor de Alencar, pursues the *bebê* during carnival in Rio de Janeiro and tries to remove the prosthesis, she resists. He complains that the pointed nose of her mask interferes with their lovemaking, yet when he finally succeeds in removing it, he discovers that it is a prosthetic designed to hide a gaping hole where her nose should be. Her face is described as "uma caveira com carne"

(75; a skull with flesh). As Julie Jones has pointed out, the *bebê* can be likened to a bride of death, an iconic figure of the *memento mori*, since she is simultaneously child, woman, and crone, encapsulating the three ages of existence in one (30). She is also ambiguously gendered, since one of the friends listening to Heitor's tale wonders if the *bebê*'s aggressive pursuit of Heitor could mean that she is actually male, thus introducing the issue of gender into the story. Notably, the word *bebê* itself is masculine, while the character is presented as female.

While João do Rio's text has generally been interpreted as reflecting the "dance of death" and the Baktinian, gothic entwining of the themes of sex and death (Jones 31), I prefer to interpret the *bebê* as a type of proto-cyborg. Although the prosthetic nose is not a high-tech gadget, it is a prosthesis that allows the character to lead a normal life, albeit only briefly and during carnival. In a sense, this primitive prosthesis functions as a baroque cyborg body part, an improvised technology of resistance and adaptation in the manner of the baroque ethos, as she survives on the margins of society. As an allegory of Brazil's social disparities at the turn of the century, her use of a façade has a parallel in the French-styled architecture of Rio's newly "sanitized" and modernized downtown city center, where the story unfolds. The couple's lovemaking takes place near the Museum of Fine Arts and the Music Conservatory and other new French-style buildings built on a massive scale, which took the place of the demolished *cortiços* (ghettos) and humble homes of the razed Morro do Castelo that once stood on the same spot. This transformation was spurred by the state's aspiration to model Rio de Janeiro after the French capital, with wide avenues and monumental government offices, along with the fine arts museum, national library, and state theater. Hence, the proto-cyborg body shares features of *codigofagia* since it contains and reveals the contradictions of Brazilian society at the turn of the century: the modernity of the newly remodeled capital, a mere façade hiding the decayed and neglected social body underneath.

By the time João do Rio writes his story in 1910, many of Latin America's capital cities had embraced European definitions of modernity. As Beckman summarizes, travel by train and steamships, "Hausmannized boulevards" or renovated urban city centers, and department stores announced entrance into a new stage of progress dreamed of by turn-of-the-century Latin American elites (141). In contrast, the cyborg or artificial body reveals death and decay as well as the unresolved issues of the masses of former slaves, soldiers, and laborers who are forced to the hills and outskirts of the city (Fischer 232–33) and who remain outside modernity and the modernized downtown streets except during carnival. As the *bebê* explains to Heitor, it is only during carnival and when she is wearing her mask that she can enjoy her sexuality. In this sense, she celebrates the baroque ethos by living in the here and now, surviving on the margins of the capitalist economy with the help of her rather crude prosthetic. Interestingly, the *bebê* does not become a literal object of sexual consumption because Heitor flees in horror before his seduction is complete. Nonetheless, she is a metaphorical object of consumption, as Heitor uses her story as entertainment to be circulated like currency among his rich friends, who "consume" her story while partaking of drink together in an elegant salon.

While neither Machado de Assis nor João do Rio brings up issues of race directly in their proto-cyborg stories, both authors are Afrodescendants and aware of the implications of slavery, race, and class and the complex system that they had to navigate in order to rise to cultural heights in Brazilian society.[10] Neither Machado nor João do Rio trusted the pseudo-scientific discourse of racial and sexual degeneration that informed the positivist-minded governmental and social elites that came to power with the declaration of the Republic in 1889. Instead, these authors employed a type of baroque ethos in their literature, using irony as a way of undermining the elite point of view and revealing the prejudices of Brazil's aristocratic class and *nouveaux riches*, who, of course, were "blind" to their critiques.

Before finishing my analysis of proto-cyborgs, I wish to mention Pedro Castera's speculative novella *Querens* (1890), since it is the only proto-cyborg text published in Mexico during this period that explicitly raises issue of race and gender.[11] Regarding the latter, the novella takes the peculiar approach of relating the female body to science based on magnetism and hypnosis. While these techniques were thought to be forms of psychological therapy at the time (Haywood

Ferreira 158–59), in Castera's tale, the woman is treated more like a machine whose systems can be tinkered with.

The story of Castera's *Querens* is centered on a beautiful young woman who was born with an impaired brain that has left her with no independent life of the mind. The woman is described as a lovely example of indigenous beauty, "una Eva pero indiana" (415; an Eve, but indigenous), which calls attention to foundational myths (Eva or Eve) of femaleness and indigenous identity, the latter feature being absent from the two Brazilian texts. Curiously, she is able to speak when hypnotized by a scientist, but she is only able to repeat his thoughts and ideas. When the scientist's colleague, a pharmacist, witnesses this phenomenon, he is enchanted by the woman's beauty and her sage words and later confesses his love for her to his friend. As Luis Cano notes, the story is more about the homosocial relationship of the two male scientists rather than a heteronormative romance (217). Although the cold scientist is interested exclusively in his experiments and has no sense of affection for the woman, he acknowledges the pharmacist's feelings, and the two men set out to "cure" her in order for the couple—the hypnotized woman and the pharmacist—to marry. However, the woman remains entirely passive and continues to speak only while under hypnosis. As the story ends, she falls into a permanent cataleptic state.[12]

In her study of the novella, Rachel Haywood Ferreira compares the woman to an automaton, quoting the scientist as saying that she is an artificial body acted upon by forces of modern technology and energy (*Emergence* 161). I find numerous hints that she might more accurately be seen as a cyborg, as when she is compared with various technologically advanced machines, for example, a "fonógrafo" (449; phonograph), a "maquinaria de un reloj" (449; clockworks), or a "camera obscura" (423). The attempt by the two men to give independent life to an inert body recalls the Frankenstein theme, although unlike Shelley's protagonist, they have no fear of animating their female creation. However, in Castera's story, it is the future "bride" who remains unreceptive to a "cure" through experimental hypnosis or "magnetism." There appears to be an interruption in the neural circuit that would complete her coming to consciousness, or the "encoding" that would allow her to express herself independently. Ironically, she is described as a burden to the scientist, who complains of the costs of maintaining her, saying, "roba mi vida para nutrirse" (447; [she] is stealing my life force to nourish herself). Thus, the burden and inscrutability that her body represents to the scientists—as both female and indigenous—casts her as a drain on scientific minds. Finally, her immobility or unresponsiveness could be interpreted as a gesture of the baroque ethos, because she appears to assent to the scientist's ideas verbally while hypnotized, but she refuses to respond to his suggestions, resisting domination through her inert body.

Ultimately, the indigenous woman's body in *Querens* has no place within the discourse of Western science, medicine, and modernity. The fact that she cannot speak for herself demonstrates how racial and sexual differences were perceived by the white male elite. Her inert body reflects the impasse of the educational and scientific elite in late nineteenth-century Mexico who, like the positivists of Brazil, failed in their efforts to "discipline" the subaltern body of the racial and sexual Other in order to incorporate it into the discourse of modernity, whose underlying principles included premises of scientific racism. The story is paradigmatic of the ways in which speculative fiction uses problematic or marked bodies to question the premises of progress, race, national identity, and modernization during this period of the late nineteenth century, when both Mexico and Brazil would struggle with these issues. It would not be until the end of the Mexican revolution in 1920 and the start of the Vargas regime in Brazil in 1930 that *mestizaje/mestiçagem* would begin to replace the nineteenth-century positivist ideology of racial whitening, as the idea of national identity based on racial mixture emerged as a platform for consensus building.

Resistance and Rebellion of Mid-Century Industrial Female Cyborgs

It is during the postwar period that Mexico and Brazil fully embrace modernity, embarking on policies that would lead to unprecedented growth, making them the most important and successful examples of late economic development. Growth rates of 6 percent per year during the period 1940 to 1975 (Graham 18) attest to this success in implementing import substitution followed by increased investment by transnational corporations (Hewlett and Weinert 1–2). This period transforms the societies of both countries as they begin to produce durable goods, including washing machines, refrigerators, cars, and telephones, in addition to more traditional products such as textiles, foodstuffs, and clothing. The female cyborg now appears as a consumer product, as both an object of consumption and a vehicle for resistance that enters the domestic space of the home. Historian Joanne Hershfield observes, interestingly, that during the import substitution period, advertisements for domestic appliances suggest "that technological artifacts are functionally an extension of the female body" (64), effectively turning them into gendered cyborgs.

The cyborg literature that emerges during this period illustrates the hopes and pitfalls of industrialization and labor, sustained by metaphors of energy and thermodynamics. I argue that these patterns of economic development are most apparent in the metaphors of homeostasis and feedback loops as embraced or rejected in the baroque ethos of mid-century cyborgs. The national ideology that officially endorses *mestizaje/mestiçagem* to affirm national identity can be extended to include a new "mixed" relationship between humans and machines.

Hayles traces how the concept of self-regulation in biological, mechanical, and economic systems grew in popularity during this period and was soon applied to information and cybernetic systems through feedback loops, based on a basic design of input → machine analysis → output → feedback → input (86). The feedback loop seems to be a particularly appropriate metaphor for Mexican and Brazilian industry because, in both cases, the state is a primary and active partner in forging economic growth. In order to promote consumption and minimize political dissent, government campaigns are mounted through print media, films, radio, and music to create a sense of national pride as a means of creating consensus to ensure the smooth running of the economy. As the governments of Mexico and Brazil work in close alliance with economic elites to produce consumer durables and other products, industries strive to create new patterns of consumption for local markets and provide consumers with desirable products for internal consumption. Hence, nationalism in ad campaigns generally aimed at the middle class was also reinforced through popular outlets of entertainment such as musical genres (*samba* and *ranchera*), sports (*futebol* and *lucha libre*), and films (*chanchadas* and *churros*). Within a social feedback loop, advertising, consumption, and entertainment work to maintain social homeostasis and guarantee the smooth running of society.

Since the postwar period is a time when national industries begin to control local production, it is an apt period for examining the transition from industries that once provided basic necessities with use-value (foodstuffs and textiles) to the production of consumer durables (appliances and machines) that constitute value-added goods.[13] In the cybernetic fictions of this period, I note that heat and sexuality seem to drive these systems and that the division between human and machine begins to fade as feedback loops connect humans to machines. Hayles traces this phenomenon back to the laws of thermodynamics and comparisons between the human body and early steam engines, which could be controlled by governors and other feedback mechanisms that were later applied to models of human behavior in literature and assembly-line culture, equating overworked humans with motors that suffered from overheating, breakdown, dissipation, and entropy.[14] In the postwar period, as Hayles documents, entropy came to be equated with randomness in cybernetic systems, and cyberneticists were able to "discount" entropy as noise or chaos that must be eliminated from the system through feedback (101). The father of cybernetics, Norbert Wiener, established the parameters that depended on the separateness of man and machine using feedback loops to self-regulate informational systems and return them to homeostasis. Such divisions avoided "overheating" or informational overload (entropy and chaos) and were programmed to return to "normalcy." Gender politics, economics, and politics are allegorized as feedback systems in cyborg-human relations in mid-century Mexico and Brazil.

The presence of female cyborgs and sexuality to be used as "energy" conduits serves to estrange us from the principles of pure profit and consumption in these developing economies. While thermodynamic metaphors may indicate a possibly unsustainable pace of annual economic growth, they may also be viewed as the "heat" of increased political pressures and social disparities. The cyborg, as both a worker and a product, exemplifies the baroque ethos of working within the system while also resisting the disparities of consumer capitalism during the import substitution period and its growing dependence on transnational capital from 1950 to 1980.

As the Mexican and Brazilian economies implement policies of import substitution to create new internal markets in the 1940s, industries have to appeal to different classes of consumers, including women. I will begin my analysis with one of the most satirical Mexican authors, Juan José Arreola, who uses the tools of speculative fiction to question economic expansion and new technology, parodying the style of advertising in his cybernetic tales. Pablo Brescia has focused his attention mainly on Arreola's "Baby H. P." a story recognized for its originality in harnessing a child's energy by storing it in a device strapped onto the child's back. As Brescia has noted, "Baby H. P." (horsepower) is basically a gadget story (93), albeit a tongue-in-cheek one that questions our concept of humanity by instrumentalizing a child or baby through a circuit or feedback loop that effectively turns it into a cyborg (101).

While "Baby H. P." is directed at mothers or female consumers, Arreola's 1961 story "Anuncio" (Advertisement) is more germane to the analysis of the "cyborgization" of the female body. In either case, whether child or woman, the body is commodified as domestic labor, either through play or sex. These stories do not depend on emotion or affect, but rather irony, since neither of them features characters or plot in a conventional sense, but rather adopt the style and content of advertisements. In reading "Baby H. P." or "Anuncio," the reader takes on the role of the scientific observer of a cybernetic system, that is, of input/text, reading/processing, and output/interpretation, in a parody mirroring the first phase of cybernetics and its discrete parts.

In Arreola's "Anuncio" (1961), female sex robots are advertised as a means of improving the lives of both men and women, albeit in different ways. Although "Anuncio" has been compared to the ancient myth of Pygmalion and the creation of the "perfect" woman,[15] I feel that it is more about the discourse that shapes attitudes toward women's bodies and experience, as well as the creation of commodity fetishes and the reification of sexual love. The opening of the story combines two biopolitical themes studied by Foucault: the institutional control of the body and the discourse and use of sexuality in modern societies.[16] This takes the form of a reference to the institutions that have traditionally been used to discipline the male body: the armed forces and prisons, in both of which women and sexual activity would be disruptive. A more efficient solution, the story suggests, is the institutional use of the female "Plastisex" robot, which provides benefits for three different civilian groups: male consumers, society at large, and women. First, male users will enjoy the ability to customize the dolls in terms of age, race, appearance, scent, and feel, with numerous options and special sexual features. Second, society will benefit from the use of Plastisex dolls through lowered levels of prostitution, which will have the effect of improving health and society. And third, women will be liberated from the burden of providing sex: "Y las mujeres, libres ya de sus obligaciones tradicionalmente eróticas, instalarán para siempre en su belleza transitoria el puro reino del espíritu" 83; And women, now freed from their traditional erotic obligations, will imbue their transitory beauty with the pure realm of the spirit). Since women will no longer have to worry about looks and erotic duties, they will be able to pursue spiritual and intellectual interests more fully, freeing them from objectification and the tyranny of the male gaze.

Notably, Arreola calls attention to the body behind the commodity fetish, modernizing and industrializing the profession of sex workers in a consumer product. He ironically proposes new networks of power, since the consumer who acquires a Plastisex doll effectively participates in a feedback loop of a cybernetic system. Foucault notes that in the nineteenth century, "biopower was without question an indispensable element in the development of capitalism; the latter would not have been possible without the controlled insertion of bodies into the machinery of production" (*History* 140–41). This idea of "insertion" is parodied in Arreola's story in a sexual way, but the story also distinguishes between femininity and sex. Sex belongs to the female body and to men—who activate it in the robot—while women have femininity. Arreola's parody is part of his use of irony as baroque resistance to modernity and its principles. While "Anuncio" purports to allay concerns with moral and religious objections to the product, as if to assure consumers that it will lead to a state of homeostasis and normalcy, it nevertheless would have the effect of transforming traditional male-female relations as well as those between human and machine.

Arreola's text also invites readers to resist the pitfalls of sexual double standards and the rhetoric of "liberation" and modernity that the advertisement appears to promote. Foucault maintains that sexuality is a product of networks of power generated by medical discourse on sex by powerful institutions, which actually forms the basis of sexuality rather than any human spontaneous expression or liberation (*History* 44–45). In Arreola's story, sexuality is openly discussed and also functions as a discourse of power, both in advertising and in its reference to two of the primary institutions of biopower, the army and prison, directed primarily at men. Arreola's use of advertising language illustrates its power to control women, and while the marketing pitch promises them "liberation," the opposite is implied.

In his style of narration, Arreola performs a type of *codigofagia* that reassembles the disparate discourses of high literature and advertising in a way that refutes capitalist logic. Arreola takes the objectification of women to an extreme in order to illustrate the traditional divide between two types of contemporary "public" women in Mexico: the prostitute and the public intellectual. The story suggests that women have no innate sexual desire, and that, when freed from this obligation, they will attend to the "pure realm" of spirituality or intellectuality. Arreola was born in 1918, a generation that also included important women writers, among them Elena Garro (1916–1998), Pita Amor (1918–2000), and Rosario Castellanos (1925–1978). As Emily Hind has documented in her study *Femmenism and the Mexican Woman Intellectual* (2010), each of these writers handled her sexuality in a different way, and as a result, was considered differently by the public. Castellanos hid her sexuality and was taken seriously as an intellectual, while Garro and Amor emphasized their femininity and sexuality, ultimately tarnishing their public images (159). This appears to show that women must maintain a state of asexuality, that is, remain in a "puro reino del espírito" (83; the pure realm of the spirit), to use the words of "Anuncio"—in order to avoid the underlying sexism in Mexican society. The baroque ethos in Arreola's tale appears in its implied resistance to an industry that would "modernize" and "liberate" sexual relations for women. The subordination of the body to the industrial model mocks Mexico's desire for new levels of productivity and industrialization.

Arreola's story illustrates what Hayles calls the first phase of cybernetics, because there is a clear division between the human male and the artificial female in the circuit and their interactions are quantifiable. In the story, the "coupling" between man and machine is not complete, since the female sex robot is an appliance that can be turned off and placed in a closet: "Es inerte o activa, locuaz o silenciosa a voluntad, y se puede guardar en el closet" (132; It can be inert or active, loquacious or silent with the flip of a switch, and it can be kept in the closet). In Arreola's text, although the human male and female robot are connected by a feedback loop, they remain separate, uniting momentarily in their heterosexual relations to "let off steam," after which they quickly return to "normalcy."

Hayles notes that this kind of closed system was challenged by the second phase of cybernetics.[17] This new phase of the paradigm is based on "reflexivity" that allows for the further integration of humans and machines.[18] With a less tightly controlled informational loop, "reflexivity" results in what Donna Haraway has called "disturbingly and pleasurably tight couplings" between humans and machines (152). The playful connotations of this quotation help in the analysis of a text by Brazilian Dinah Silveira de Queiroz, whose female protagonist initiates a relationship with a cyborg in which the sex is less mechanical than in Arreola's text, eventually turning into a mutual exchange of information that is reflexive and cybernetic in nature.

While both Arreola's and Queiroz's texts rework the idea of labor, robots, and sex as economic energy, and both participate in feedback loops that resist the discourse of modernity in uncanny ways, Queiroz's "O Carioca" (1960) also incorporates more of the baroque ethos in the behavior between cyborg and human. Queiroz's female protagonist resists the strict profit use of the cybernetic creation and effectively introduces "chaos" into the system. The story begins as two unnamed neighbors, a man and a woman, become friends when he is taken ill. She later learns that he has invented a cyborg he calls "the Carioca" as well as others that are his cybernetic pets. Later, the couple enters into an awkward intimacy, which she finds unsatisfying. One day, while the man is away working, she enters his apartment to take care of domestic chores and ends up enticing the cyborg into a cat-and-mouse romantic game that the man himself had refused to engage in.[19] She does this by breathing on the machine's "neck" in a way that satisfies its desire for heat, creating a feedback loop that resembles sexual love play. She regularly visits and "overheats" her partner only to be found out by the man, who promptly dismantles the machine, not just out of jealousy, but because, as he claims, she has ruined the machine's programming for military use.

Commenting on "O Carioca" in 1963, David L. Dunbar emphasizes the "robotic" personality of the male neighbor (126), who appears even less human than his artificial creation. While this appears to be an amusing or humorous touch, Luciana Monteiro emphasizes the proto-feminism in the woman's attitudes when she takes on a provocative role with o Carioca (731). I would note that she also continues to perform traditional wifely duties, doing the neighbor's housekeeping, taking care of his cyborg pets, and nursing him back to health. While she does not completely break gender stereotypes, her attitude recalls the baroque ethos because she resists them, repurposing the cyborg for a spontaneous erotic encounter that is beyond its original purview. She finds "use-value" or domestic consumption for the robot, whereas the man had planned to sell it to the Americans for its "exchange value" or profit. Unbeknownst to the man, Queiroz's female character is also able to elicit affection from the man's other pet-like cyborgs, a pair of foxes, a turtle, and a monkey, all of which cuddle with her. This story captures Echeverría's baroque ethos in that the woman's sexuality—as well as her maternal instinct in the case of the cyborg pets—resists capitalist uses of technology and reconfigures relations for domestic or internal consumption. Her relationship with the animals resembles ideas first outlined in Haraway's

"Cyborg Manifesto," in which humans acknowledge their intercon-
nectedness to other "life" forms, treating them as equals instead of
as objects of exploitation (154).

By the end of "O Carioca," the cybernetic reflexivity is complete.
When an attaché sent to pack up the man's remaining possessions
asks the woman to return the key to the man's apartment, she freezes:
"O coração dela diminuiu. . . . Mas reagiu e seu motorzinho entrou
a funcionar: devia ser um engenho que se comportava segundo o
modelo preestabelecido" (203; Her heart shrank. . . . But it reacted
and started up its little motor: it was like a device that behaved in
accordance with a pre-established program). In short, the woman
describes herself like a cyborg running a program that can be "reset."

Finally, I would argue that the fact the man preferred selling his
product to the American military to allowing its use as a playmate
reflects the period's politics. The woman's "heating up" the situation
not only illustrates tensions of gender politics by introducing "chaos"
into the system, but also brings up issues related to emerging political tensions. If the cyborg represents Brazil's body politic, its fate also
appears to foreshadow the dismantling of the Brazilian democratic
state by an elite that does not want to share power. By the end of
Queiroz's story, the woman realizes that the cyborg, whom she vowed
never to forget, will likely be submitted to a "lavagem de cérebro" (204;
brainwashing), the fate of Brazil's fledgling democracy. It is proba-
bly not a coincidence that the man in the story works for the armed
forces and awaits a military jet to take him to his next location, while
the woman goes out into the streets to take a bus among her fellow
citizens. As they go their separate ways, the tensions between male
and female, technocrat and citizen, authoritarian rule and democratic rule have been exposed but remain unresolved and tamped down.
The story appears to foreshadow some twenty years of military rule
(1964 to 1985) in Brazil, which attempted to re-establish high levels of
economic growth and bring about a return to traditional values of
church and state, that is, the "normalcy" of the 1950s class structure
and traditional elite privilege.

In conjunction with these two stories above, I would also like to
briefly mention Alfredo Cardona Peña's "La niña de Cambridge" (1966;

The girl from Cambridge), a parable about a female cyborg—a version 2.0 working in the presence of her "mother," version 1.0—who is asked to perform an unsolvable calculation by male scientists. When the impossible task makes her explode, the scientists who fed in the request are charged with cruelty to thinking machines. Perhaps this female cyborg is meant to represent all women: as the young male scientists surround her, feeding in more and more questions and causing her to overheat, the story begins to read like a gang rape, where the only solution for the young female cyborg is to disappear, sublimating her violation. The mother cyborg subsequently loses her rational computing capacity, repressing the crime and instead interpreting the event as a manifestation of the divine, that is, the Virgin's divine assumption. Ironically, Cardona Peña's story suggests that at times it is easier to be asexual or even immortal than it is to be female. This again illustrates Echeverría's idea of the baroque ethos; although the female cyborgs appear to assent to their trials, their reactions to abuse by disappearing or going insane attest to their sense of loss.

Cyborgs continue to serve as an apt vehicle for critique and gender politics in the following decades as both countries faced increasing resistance to political repression. These pressures came to a head in 1968, when both Mexico and Brazil were faced with student protests, leftist kidnappings, and bank robberies that both reflected and exacerbated the increasingly authoritarian stances of the respective governments. The year 1968 marks the coup within the coup in Brazil, when military hardliners cracked down on leftists and all forms of artistic and journalistic forms of protest. In Mexico, 1968 is also the year of the government's massacre of student protesters in Tlatelolco on the eve of the Olympics, an event that rocked Mexican society and was effectively censored by authorities.

As agents of political crisis or victims of social upheaval, female cyborgs simultaneously contest gender expectations through action as they express the contradictory attitudes working within the system in a way typical of the baroque ethos. As they move from more conventional feedback loops to cybernetic reflexivity and receptiveness to the Other, we see how authors begin to use science fiction paradigms to portray Mexican and Brazilian society.

Moving on to the 1970s, I begin with a story by Brazilian Caio Fernando Abreu, a mainstream writer who is best known for his themes of bisexuality and queerness (Arenas 13–14), typified by his story "A ascensão e queda de Robhéa, manequim e robô," (1975), about a robot/cyborg endowed with gender. In this story about the Brazilian military regime and its persecution of artists and militants, Abreu uses the artificial nature of cyborgs to call attention to the exclusion, fear, and attraction of the Other in both political and sexual senses. Despite its desire for love and acceptance, Abreu's gendered cyborg is constructed as a symbol of political resistance and exemplifies the baroque ethos.

Set in a near-future society run by an authoritarian regime, Abreu's story fits into the trend of dystopian texts written in Brazil during the 1970s and '80s, when mainstream writers turned to this subgenre of science fiction in order to capture the tensions of the new industrial order imposed by the military dictatorship (Ginway, *Brazilian* 89–136). In Abreu's story, when a sudden plague affecting only humanoid robots emerges, the government denies them treatment and waits for them to "die." Cynically, the government later launches a campaign to promote the sale of their metal remains as fashion accessories, thus commercializing the diseased parts of this population and converting them into desirable objects as commodified "outsider" art. The decorative use of the cyborg body parts shows how the recirculation of goods can result in ever more subtle forms of control, since the now prestigious articles worn on the bodies of humans reveal the consumerist mindset of modernity. This circulation is comparable to feedback loops as described above, but here the robot/cyborg "outsiders" supply input for the system controlled and regulated by consumer "insiders." Thus, the capitalist recirculation of formerly taboo objects allows the regime to re-absorb the energy of resistance to maintain its own power, since excess energy (dissent) is rechanneled and allowed to escape (as fashion), guaranteeing the smooth running of the system through a type of homeostatic control of first-phase cybernetics.

Despite government censorship of "robot" culture, a journalist begins to publish news stories about the plague, turning a forbidden topic—robot/cyborg culture—into a new and different fashion trend, because humans now begin to imitate the cybernetic look by wearing metallic clothing and artificial eyewear, deepening their obsession with the forbidden Other. The journalist eventually establishes an artistic movement that attracts fashion designers, artists, writers, actors, and film directors, who raise the country's artistic status to a new international level of visibility. As Idelber Avelar ("Revisions" 187–88) notes, it becomes clear that Abreu's robots are allegories for queerness in Brazilian society, showing how this community's artistic contributions are coopted and exploited as part of Brazil's cultural identity. Again, cultural resistance is converted into cultural capital, this time on an international scale.

The end of the story finally introduces us to Robhéa, the title character. She is the unexplained "offspring" of a secret group of surviving robots and the central figure of the story. Abreu's artificial humans have families and human-like relationships and display sexual desire.[20] Despite the accepted robot fashion trends, actual robots continue to be considered subversive, with the result that Robhéa is imprisoned by the regime shortly after being discovered. Later, she is released from prison and sponsored by a fashion designer who helps her become a famous model and actress. However, after five years and at the height of her success as an icon of Brazilian *carnaval*, amid rumors of her homosexuality, she isolates herself on an island. When a bestselling tell-all book cataloguing Robhéa's lesbian exploits is published, she commits suicide. In a final irony, the regime turns even this most extreme form of protest to its advantage by erecting a memorial in her honor, continuing the policy of cooptation of subaltern protest by the Brazilian state.[21]

In his analysis of Abreu's story, Avelar points out that the author's use of the robot's sexual orientation and gender identity helps us understand attitudes toward transvestites and gay culture in Brazilian society ("Revisions" 187). It could be argued that the portrayal of robot culture in the story portrays the uncanny combination of acceptance and rejection of queerness in Brazilian culture, since while it is accepted in carnival and melodrama, where gender performativity and high levels of emotion resonate strongly (H. Perez n.p.), it is rejected on a more basic level, as seen in violence against LGBTQ communities.[22] This shows that the body and sexuality remain central to the construction of the Latin American cyborg. Abreu's robots manage to form colonies, resist, reproduce, and survive, and while their unstable identity provokes anxiety in and persecution by humans, they are subsequently used to generate wealth and cultural prestige. In her examination of the posthuman body in *El monstruo como máquina de guerra* (2017), Mabel Moraña concludes that the cyborg body escapes classification in a way that implies contagion (235), recalling the fear of the spread of the robot plague at the beginning of the story. Abreu anticipates gender conflict by highlighting the unstable cyborg characteristics of Robhéa, who is simultaneously heroic and fragile, hardened and vulnerable as an artificial female who threatens the foundations of society.

In some ways, Robhéa's struggle calls attention to a baroque form of resistance that is distinct from that of the masculinist heroic narrative. In his study of the literature of political resistance, Idelber Avelar points out that any attitude other than open defiance is framed as suspect, "feminized," or "silenced" (*Untimely* 66). I argue that it is the gendered nature of Robhéa's struggle that avoids the heroic rhetoric of realist texts associated with the dictatorship period as analyzed by Avelar, who describes how autobiographical testimonial narratives were used to codify political resistance and to suppress issues of loss, mourning, and defeat in national consciousness (*Untimely* 62).

Notably, Robhéa's story appears in the 1975 anthology *O ovo apunhalado* (The stabbed egg), suggesting the ambivalence of Abreu's vision: the egg with its hard shell represents a masculinity that contains the fluids associated with femininity and reproduction. Using the image of the stabbed egg, Abreu evokes the violent rupture of the symbolic order in his own ambiguously gendered way, that is, by depicting the collapse of the border between the "inside and outside," masculine and feminine, threatening traditional heteronormative concepts (Kristeva 232). By breaking taboos among literary genres (mainstream and science fiction) and by using queer characters that straddle sexual gender, Abreu uses the hard robot shell to contain the vulnerability of Robhéa inside. Her trajectory, through pariah status, success, and finally cooptation mirrors the experience of outsider artists in Brazilian society in the 1970s. Finally, one might argue that the uncanniest aspect of this story—written in 1975—is its anticipation of the AIDS epidemic.[23] The portrayal of cyborgs who long for acceptance and community captures the tensions between tradition and modernity in a baroque ethos of adaptation, survival, and resistance within Brazilian society.

Abreu's writing is radical for the period because he avoids facile divisions between robots and humans, homosexuality and heteronormativity, resistance and collaboration. He critiques military repression, but does not turn any of his characters into heroic figures of resistance typical of the realist *romance-reportagem* (journalistic novels) of the period. Avelar argues that such narratives allowed the Brazilian public to experience a sense of catharsis and relief without guilt or mourning (*Untimely* 61–63). Flora Süssekind makes a similar point, noting that these texts, although critical of the regime, fit into the naturalist paradigms of Brazilian culture and society, whose "growing pains" were part of the country's long road to the "utopia" of a modern nation (57).[24] Thus, after 1985 and the end of the regime, the rhetoric of democracy was overlaid onto a society whose institutions remained largely unchanged. Instead of exorcising the violence of the past, its civilian leaders simply pushed it away while steering toward a neoliberal future of global competition.[25]

A gendered cyborg also figures prominently in Emiliano González's 1978 story "Rudisbroeck o los autómatas" (Rudisbroeck or the automatons), and although it is by no means a feminist tale, it has much to say about cybernetic reflexivity. It is also characterized by Borgesian themes of infinite regression and a baroque ethos.[26] As the story of a rebellious female automaton who, by disobeying her creator, brings down a curse on her people, it appears at first to be akin to the betrayal of La Malinche, "the impure, the flawed, the opportunist" (Messinger Cypess 16), but it soon becomes apparent that the story is a commentary on such myths in Mexican society. The nineteenth-century feel of González's story pays homage to the horror influences of the likes of Arthur Machen and H. P. Lovecraft and the kinds of pulp fiction that appeared in magazines such as *Weird Tales*. While clearly more a work of fantasy than science fiction, González's story is baroque in both its digressive form and its content, which is resistant to the paradigms of modernity.

The story takes place in the eerie sepia-colored world of Penumbria where, paradoxically, time is frozen, but events are not. An unnamed narrator meets an inventor named Rudisbroeck who has created a female automaton to take the place of his lover, the adolescent Glinda, so that he and Glinda can run off together unbeknownst to Glinda's mother, a fairy queen. This plan is foiled when the female robot—who is strikingly human—kills the real Glinda in a fit of jealousy. Rudisbroeck forgives his creation but fears the revenge of the fairy queen, whose curse has frozen time in Penumbria—much like the curse of the witch in Sleeping Beauty, whence comes the title of González's 1978 collection in which this story appears, *Los sueños de la bella durmiente* (The dreams of sleeping beauty).

It is significant that, rather than punishing his creation for killing his true love, Rudisbroeck instead recognizes her humanity. Gone are any manifestations of feedback loops and first-stage cybernetics, and only the second stage of reflexivity and the uncanny humanity of the automaton/cyborg remain. The interchangeability of the actual and artificial human confirms Hayles's concept of "reflexivity," blurring the conventional boundaries of inside and outside, of human and posthuman. In explaining the pitfalls of reflexivity, Hayles cites the infinite regression in Borges's 1941 story "Las ruinas circulares" ("The Circular Ruins"), in which a narrator creates a student through dreaming, only to discover that he is part of a dream of another, who in turn is part of another, and so on (*Posthuman* 8–9). In González's story, it appears that the narrator has become part of Penumbria and its collective dreaming. Rudisbroeck and Penumbria are real, but both Glinda and her mother's curse turn out to be part of a foundational tale that the inhabitants of Penumbria are collectively dreaming.

González calls attention to the fact that Mexico, with its many historical myths, has something in common with the dream-like state of Penumbria. Brian L. Price labels this telling and retelling of Mexico's past as the "cult of defeat," or the obsession with its suffering and historical impasse, reformulating it into a transcendental pact with

martyrdom (4). In analyzing González's *Los sueños de la bella durmiente*, Serrato Córdova considers González's recourse to fantasy to be
a protest against the modernizing exhortations of Luis Echeverría's
presidency (1970–1976) and its slogans of self-sufficiency and promises of unprecedented wealth based on the oil industry (37). In this
sense, Emiliano González's 1978 cyborg tale is also baroque in its use
of automatons (technology) that resist the official history of progress and nation building in modern Mexico.[27]

González's counternarrative of progress and the deliberately outmoded style of his work also recall a passage by Brazilian anthropologist Roberto DaMatta, who pointed out that observations about Latin
America require a special optic: "one speaks here of another 'American' reality, one without social linearity, where time seems to have
passed without generating 'history'" ("For an Anthropology" 271).[28]
Perhaps instead of considering González's work to be a rhetoric of
defeat, we might understand it as illustrative of the baroque ethos in
its resistance to modernity and its inclusion of humanized cyborgs
and reflexivity. It also marks the transition from the first stage to the
second stage of cybernetics, heralding the neoliberal age.

Neoliberal Cyborgs and Gender

It is at this point, the transition from the second to the third stage of
cybernetics, that our relationship to machines and digital interfaces
changes radically. Cyborg stories in the mid-1980s mark the advent of
the personal computer and what Hayles calls the posthuman, when
the boundaries between human and machine all but disappear. She
points out that the flashing cursor becomes a metonym of Jacques
Derrida's concept of presence/absence, life/death, or pattern/
randomness typical of binary digital cybernetic systems (43).[29] In
this phase of cybernetics, the notions of presence and absence are
inverted: the "presence" of the body loses out to the "absence" of
virtual space, that is, "cyberspace defines a regime of representation
within which the pattern is the essential reality and presence an optical
illusion" (36). In his well-known 1984 essay "Postmodernism, or the
Cultural Logic of Late Capitalism," Fredric Jameson has claimed that

the new information society is the purest form of capitalism, since bodies, products, and entertainment in digital form can be more easily traded, bought, and sold (77–80). This also facilitates the implementation of neoliberal policies and privatization, as disembodied data travels across global economic systems.

Historically, the 1980s are a time of crisis in Latin America, and Mexico and Brazil are no exceptions. When devaluations left the Mexican economy in shambles in the 1980s, President Miguel de la Madrid reacted with austerity measures to satisfy international bankers (Lopez-Lozano 35). The situation was worsened by the General Agreement on Trade and Tariffs (GATT), which took effect in 1986, ending import-substitution, and by a new immigration act that increased tensions for temporary workers in US border states (Skidmore and Smith 250). In Brazil, the year 1985 is associated with the end of the dictatorship, which, despite its repressive measures, protected key national industries and interests. With the opening of new global markets, Brazilian workers began to feel the pressure of competition, primarily in the form of job losses. Throughout the 1990s, much of the population suffered under hyperinflation and other economic hardships that marked the civilian governments of José Sarney, Collor de Melo, and Itamar Franco, in a period known as Brazil's "lost decade." Thus, neoliberal policies coincide with the emergence of personal computers and the privatization of public space and public corporations in Latin America, effectively ending the period of growth of the national economies in both Mexico and Brazil.

British economist Colin Lewis identifies the hyperinflation of this period as yet another form of violence inflicted on the general population of most Latin American countries, adding to the traumas of repressive regimes and the rise of crime in urban areas. All this contributed to the further breakdown of civil society and of the pact between the state and its citizens (39). The state's implementation of neoliberal measures and policies of fiscal discipline to halt inflation meant that much of the population suffered under the full brunt of austerity measures together with bouts of inflation. By the 1990s, new ideologies of deregulation and privatization led to the sale of state enterprises to private consortia as new trade agreements emerged

affecting both Mexico and Brazil, respectively NAFTA and Mercosur/
Mercosul. However, another important phenomenon that specifically
affected Mexico and Brazil was the rise in organized crime associated
with the drug trade, which resulted in the creation of what might be
considered a parallel state.[30]

In William Gibson's 1984 *Neuromancer*, drugs are touted as a way
of sharpening the mind of the hacker who seeks to enter secured
cyberspace. They also cut off the hacker from others, creating a private paradise and a type of cyberspace high or elation. Hayles notes
that the mind, when conceived of as information, leads to a conception of the body as a prosthesis, thus transforming bodies into storage space for minds (*Posthuman* 288). The idea of the privatization of
experience and the body as "shell" or storage unit reinforce the political alienation of the period. The connections between the neoliberal
economic crisis, organized crime, virtual reality, and drug-induced
states begin to figure in Latin American cyberpunk in the 1990s.

In cyberpunk, the imprisoned, implanted, or drugged body is used
as a social marker to signal a shift in public policies in the new global
market. Similarly, the early twentieth-century modernist dreams or
"export reveries" as described by Beckman re-emerge in the form of
digital dreams or drug-induced states described in Latin America
cyberpunk.[31] Beckman notes that illegal drugs, especially cocaine,
become the new extractivist product that supplies wealth to the region,
following routes in Colombia traveled by the protagonist of Rivera's
La vorágine (1925) in the rubber trade. She notes that, ironically, these
"are the same as those traversed today by drug runners, paramilitar-
ies, leftist guerrillas, Columbian soldiers and US Marines fighting for
control over cocaine production" (190). Thus, trails originally forged
for export commodities are repurposed for the drug trade.

To illustrate the dizzying conflation of neural networks, digital culture, and illegal drugs that appears in Mexican and Brazilian science
fiction of the 1990s, I briefly wish to cite Mexico's first cyberpunk novel,
Gerardo Horacio Porcayo's *La primera calle de la soledad* (1993; The first
street of solitude), a work that has been thoroughly studied by Hernán
García and Juan Muñoz Zapata, among others. I see a parallel between
this novel and Ivanir Calado's "O altar dos nossos corações" (1993; The

altar of our hearts), which I have analyzed as an allegory of Brazil's body politic as hijacked by a narco-state ("The Body" 201–2). Both tales involve computer viruses or implants that, much like cocaine, induce enhanced physical states and dreams. These are sold to the public and to politicians, making more explicit the connections among governmental corruption, prosthetics, digital culture, and drugs.[32] Freed from embodiment, information can travel without the constraints of time and space, unfettered by the rules of the "mortal world" (Hayles 13). While access to digital networks, trade, and information are recurrent themes of Latin American cyberpunk, bodies still find their way into the fray, never dissolving entirely into the digital realm.

Mexican and Brazilian cyberpunk/narco-narratives tend to focus on homosocial worlds of crime, corporate espionage, hacking, and politics. I am more interested in the baroque ethos of resistance represented by the female or gendered cyborg body that emerges in this period. Mabel Moraña considers the cyborg body to be one of the monsters of philosophy, a hybrid signaling the collapse of the social order or a grotesque distortion of the postmodern collective body politic (224). I believe that the cyborg body in Mexican and Brazilian science fiction can also illustrate a baroque resistance to the exigencies of the market, a protest against the disappearance of labor, the loss of the social pact between the state and society, and the new paradigm of privatization. Many of the cyborgs in these stories recall the crises of uneven modernization, political struggles, and drug wars in both countries in the 1990s. They embody the crisis of the body politic, since technology is inserted into people's bodies to restrain, subjugate, and recolonize them in traumatic and violent ways. As we shall see, in Mexico, the ambivalent, penetrated body finds itself infantilized and over-disciplined by a new corporate parental order. The protagonists of Mexican cyberpunk evoke both Kristeva's abjection and Echeverría's *ethos barroco* in their attempts to both resist and survive within capitalism.[33]

As a prelude to wider discussions of the shifting interfaces of data, bodies, and corporate power, I note that in stories such as Gerardo Sifuentes and Bernardo Fernández's "(e)" (1998) and Pepe Rojo's "Ruido gris" (1996), the worker/protagonists do not own their prosthetics, which in any case benefit only the corporation and its investors. Corporations maintain schedules such that workers have little or no contact with parents, families, or friends, and they are generally forbidden by corporate policy to use drugs or alcohol.[34] With few jobs available, the market provides a seemingly unlimited supply of workers, emphasizing the corporation's new biopolitical role where most forms of resistance are of limited effect if not entirely futile, and corporations are able to recruit a steady flow of fresh bodies for the new digital labor force.

The conflict between digital and embodied labor is taken up in depth in Bef's masterful *Gel Azul*, which received the 2007 Ignotus award for best short science fiction novel in Spanish. Most accurately classified as a Mexican cyberpunk novel, it is concerned with trafficking networks, political corruption, class divisions, and violence in the urban milieu. Sara Anne Potter points out that Bef's novel draws on both the cyberpunk world of *Neuromancer* and the detective fiction or *novela negra* of Paco Ignacio Taibo II to construct its own vision of a futuristic Mexican world of cyber-crime and politics (148). However, since Bef dedicates the novel to José Luis Zárate, I believe it is also likely that he is paying homage to Zárate's award winning story "El viajero" (1987), a story about a detective who finds that his reality is modified substantially each time a time-traveling client visits him. In the end, like Oedipus, he ends up realizing that he himself committed the crime that he is supposed to be investigating. In Bef's novel, although the detective Crajales is not literally the criminal, he does end up murdering a hacker, his former partner and best friend, Salgado, who functions as a double of himself.

In her sophisticated analysis of *Gel azul*, Sara Anne Potter focuses principally on virtual reality as it applies to the main female character, Gloria Cubil, a privileged young woman who replaces her addiction to drugs with surfing the web through a variety of avatars. I will focus more on the issues of gender ambiguity and embodiment, and the bending of gender in two main characters: Gloria and Crajales, the private detective. The novel opens with a monologue by Gloria, as she endlessly surfs through cyberspace in a tank of blue gel where all her physical needs are taken care of. It turns out, however, that even in this protected environment her body is vulnerable to attack, as she has been unknowingly impregnated.[35] Only when a workman comes to clean the tank is this central crime of the novel discovered. Crajales is hired by a lawyer, Beltrán, to investigate, but it soon becomes clear that Beltrán has other motives.

In *Gel azul*, the divisions between mind and body and male and female gender are sharply drawn at first, yet they slowly break down as the novel progresses. While in the beginning it appears that Gloria's body is more of symbol of rivalry in the homosocial world of two adversaries who hope to "win" her, Beltrán and Crajales, it is eventually revealed that the violation of her body is primarily meant as a warning to Gloria's father from men wanting to take over his share of certain black-market activities. When Beltrán and Crajales end up in a fight to the death, a typical element of the homosocial plot, their rivalry is not to win Gloria as a mate, but rather to garner the glory or power that her name represents to each. Crajales, as it turns out, wants to have Gloria's access to cyberspace, while Beltrán wants her father's power and increased access to trafficking networks for prosthetic limbs that have been surreptitiously amputated from unsuspecting victims in gel tanks.[36]

Like *Neuromancer*'s protagonist, Crajales has been "burned"— barred from ever returning to cyberspace—due to a conviction for hacking. Indeed this explains why he becomes a detective. Without internet access, Crajales has to call in a favor from his former hacking coconspirator and friend, a journalist named Salgado, whom he has to meet in person since he is barred from the web. Completely interfaced in his own gel tank, Salgado protests at first but slowly begins to disconnect the elaborate apparatus in order to meet face to face. When Crajales arrives, to his horror and that of Salgado, they discover that Salgado's limbs have been "harvested." After both men recover from their initial shock, they surmise that the crime must have been carried out by the same people involved in Gloria's case. As Potter points out, unlike Marshall McLuhan's figurative "auto-amputation," a phrase used to describe the social withdrawal of those seeking refuge in electronic or digital nirvanas, "The amputations and prosthetics in this novel, then, are real (physical and not technological) and organic" (160). This again suggests that the human body is of primary importance, and that cyberspace is used as a trap in this global war to harvest limbs for sophisticated life-like prosthetics. Salgado finally confesses that he was the one who burned Crajales, but he pays a high price for this confession when Crajales kills him in a rage.

The theme of prosthetics and amputation also brings up gender anxiety and instability in the human/cybernetic splice in the novel. When we first meet the two seemingly opposite characters—the unkempt, unattractive, masculine Crajales, a violent man of action, and the lovely, feminine Gloria Cubil, the floating Madonna, immobilized in gel—they appear to be worlds apart. Her interface allows her to live in an apparently postgender and posthuman world, with the ability to experience it in the guise of different types of beings, animals, sea creatures, and even insects. In anonymous inner monologues, one of the characters describes a desire to surf the web as an orca, an experience similar to Gloria's reveries. In the end, however, it becomes clear that the interior monologues belong to none other than Crajales, because after being hunted down by Beltrán, Crajales trades his silence on Beltrán's crimes in exchange for few final hours in cyberspace—which he will spend as an orca. However, as Hayles points out, in a data-based world, power is not about possessions but about access to information and cyberspace (*Posthuman* 39), and Crajales's access is fleeting. He only just manages to provide Gloria's father with key information on the crimes before his limbs are harvested and he is killed in his own tank of blue gel.

The baroque aspect of Bef's text is Crajales's return to the gel tank, defying the realist ethos of profit. Although the gel tank is ultimately a womb-like space associated with physical, political, and psychological amputation, it nonetheless becomes Crajales's act of defiance. The tank breaks down traditional oppositions between inside/outside, male/female, water/air, and life/death, in a type of assemblage or *codigofagia* that we have not seen in other works. Simultaneously imprisoned while feeling physically free in the gel tank, Crajales becomes a killer whale that will be soon be killed. The orca is a composite of opposites, with a body that is black and white, sleek yet dangerous, making Crajales's avatar ambiguously gendered in a curious way. Teeth, amputation, and gel all suggest the monstrous feminine, uniting the fear of castration with gender anxiety (Creed 82–83). Yet in a gesture typical of the baroque ethos, Crajales has planned his revenge in a random act to reveal Beltrán's crime in a document that disrupts the pattern of the system. Although he cannot defeat the forces against him, he can introduce chaos into the system.

The novel also addresses the trauma of the body politic. Even though this is a world where borders and nations no longer matter, Mexico still appears to be a supplier of primary resources—in this case limbs—and still inhabits a place on the periphery of the military-industrial complex. As Potter states, the blue gel is the color associated with the conservative Partido Acción Nacional (PAN), which put its own set of free-market policies into effect in the year 2000 (163). The gel, meanwhile, symbolizes a political impasse or stagnation of the system. What we see in the plot of the novel is a distortion and disarticulation of the body, of a body politic immobilized in blue gel that is penetrated or violated by global capital. The focus on the missing limbs also recalls that much of the US depends on Mexican workers, their arms and legs, for labor-intensive work. Cyberspace, as portrayed in *Gel azul*, becomes another drug for numbing the pain of societal violence. It is highly addictive, is affordable only to the economic elites, and infantilizes its users, placing them, immobilized, in womb-like tanks, cut off from the body politic.

The implied "wetware" of the flesh in cybernetic fiction, sexuality and reproduction recalls Kristeva's concept of the female body, particularly its "disowned" or "banished" nature. Moraña posits that the gendered cyborg reflects the nature of capital and the breakdown of gender and binaries, citing Judith Butler's claim that stylization and performance give gender substance (228), as seen in Bef's novel. Ultimately, Moraña concludes that the monster and women represent difference and Otherness outside logocentric culture, considering the cyborg body as a type of monster that escapes classification (235). In the cybernetic fictions penned by Brazilians Roberto de Sousa Causo and J. P. Cuenca, it is also female cyborgs who refute or question societal expectations, at times undermining heteronormativity or traditional family structures in order to resist and survive in capitalist modernity according to the baroque ethos.

The defiant female cyborg Shiroma first appears in Roberto de Sousa Causo's 2008 "Rosas Brancas," whose story may be seen as an allegory of Brazil's more recent past: she is the daughter of a soldier (suggesting the military regime), who is sold to a crime syndicate by her mysterious father (a neoliberal transaction) where she is forced to participate in organized crime (violence and narco- and human trafficking). The conflict between Shiroma's cyborg status and her human emotions can be explained in part by Hayles's concept of cybernetic "reflexivity," or the human-like presence in artificial life forms posited by Philip K. Dick (175), whose androids, as in his 1968 novel *Do Androids Dream of Electric Sheep*, are often more emotionally expressive than their human counterparts. In contrast, Shiroma's father, an impassive scientist named Perseu Sunne, has no feelings for the cybernetic beings he has fabricated in his military lab. His first creation, a mixed-race cyborg named Mara, is unaware that she is not completely human, because she experiences and displays human emotions. After serving in a military unit off-world, Mara returns to Earth, where she becomes Sunne's lover and inexplicably conceives a daughter (Shiroma). This incestuous situation, where Mara is both creation/daughter and lover/wife of the scientist, could be construed as representing the corruption of a dictatorial past that remains entwined in Brazil's current body politic. This becomes part of the heritage that Shiroma must face as she struggles to survive in her crime-ridden society.

In "Rosas Brancas," Mara finds out about her own cyborg identity when she tries to recover five-year-old Shiroma from kidnappers. It turns out that Sunne had secretly decided to sell them both to an organized crime network but ordered the transaction to be staged as a kidnapping in order to hide the sale from the military lab, which plans to stop the project because Shiroma suffers from a genetic defect. Mara proves to be more "human" than her creator, for her maternal instincts induce her to sacrifice herself for her daughter. Shiroma survives the ordeal, only to be raised by a couple who use the resources of a criminal organization to cure her and train her as an assassin. Shiroma is essentially enslaved to her new masters in a situation typical of human trafficking.

The fast-paced story "O novo protótipo," published in 2009, continues Shiroma's tale some twelve years later, now set in the ironically named neighborhood of Liberdade (Liberty), the first and principal Japanese community of São Paulo. Shiroma, who can pass as a native of this neighborhood, is now seventeen years old and is on her first mission, an assassination. She mentions in passing that she has been sexually initiated by a man and a woman before making her first kill. While she does not describe herself per se as bisexual, she admits to having sexual feelings as she recalls this first experience. However, she does not use her sexuality in her work as an assassin, preferring to use her physical prowess to defeat adversaries. As the story unfolds and she confronts the criminal who is her target, we gain insight into her psychological fragility, as she longs for her mother's advice even as she scales walls and breaks out of powerful restraints before killing her victim, an organ trafficker.

As the protagonist of a bildungsroman of sorts, Shiroma's character develops further in the ten stories of the collection *Shiroma, matadora ciborgue* (2015). In the last of these stories, "Tempestade solar" (Solar storm), which was actually published separately in 2010, Shiroma ends up killing Sunne as well as her criminal captors, an act that recalls Haraway's dictum that cyborgs are seldom loyal to their father-creators. As a part Asian, part mixed-race, sexually queer cyborg assassin, Shiroma combines genetic and moral codes in a way reminiscent of Echeverría's *codigofagia*, constructing herself as a unique cultural assemblage. She resists both the military, as represented by her father, and her organized crime bosses, while longing for her mother. Despite her work as an assassin, Shiroma is a sympathetic character whose search for survival allows her to work within a system while simultaneously attempting to escape it.

Shiroma may also be regarded as an appropriation of the cyborg assassin figure that first appeared in Japanese cyberpunk manga and animé, most famously in the 1995 film *The Ghost in the Shell*, in which a young woman's brain is put into a prosthetic body—hence the idea of a ghost (her spirit/brain) in a shell (body) as explained by Sharalyn Orbaugh (445). Yet Causo's Brazilian version—Shiroma—differs from the Japanese cyborg because she was conceived and born in a conventional way, retaining a mostly human body and strong ties to her mother, whose voice she hears through a mysterious seashell that acts as an interface between them.

Shiroma often deploys her apparent physical fragility to gain the upper hand on her missions, since her delicate appearance disguises her unexpected masculine strength. Given her connection with her mother and her political function as a *justiceira* or equalizer in a society riddled with crime, she presents a different configuration of cyborg: she does not present the mind/body split of the Japanese series (since she retains human organs and skin) nor the independent traits of the Harawayesque postgendered cyborg (in her continued attachment to her mother). Causo's story presents an unlikely illustration of the *ethos barroco* in its reverence for motherhood (tradition, conservatism), the struggle for independence from organized crime (resistance, inconformity), and Shiroma's mission of social justice.

In her *Embodying the Monster: Encounters with the Vulnerable Self* (2002), Margrit Shildrick asserts that the vulnerable or monstrous body brings out "the otherness of possible worlds, or possible versions of ourselves not yet realized" (129). This suggests that the cyborg could represent a form of existence between the natural and the technological, offering a sense of emancipation. It frees us, if only momentarily, from biological, cultural, and technological constraints in a utopian alternative implied in Shiroma's search for freedom, identity, and community.

This utopian possibility is also brought up by the sexualized female posthuman of J. P. Cuenca's 2010 *O único final feliz para uma história de amor é um acidente* (*The Only Happy Ending for a Love Story is an Accident*).[37] Set in Tokyo, the story begins with a "female" android named Yoshiko who begins life as a sex doll made to the specifications of an aged Japanese poet, Atsuo Okuda. She narrates several chapters in the first person in the novel, and while she initially appears to be a minor character, I believe she holds the key the story as the representative of resistance and survival of the baroque ethos.

In this novel, the city of Tokyo is portrayed as a giant cyborg whose eyes are the cameras that monitor public places, recalling Foucault's description of the Panopticon and surveillance in modern life

(*Discipline* 201–3). While the city (and patriarchal society) are "male," the female cyborg (in the figure of Yoshiko) belongs to private space and appears to be confined and powerless. Caught between these forces is the poet's son, Shunsuke, who initially does not realize that his father has hidden cameras in his apartment and has engaged a group of informants to keep him abreast of his son's activities.

Okuda disapproves of his son's relationship with his new lover, Iulana, a Polish-Romanian woman, and threatens to kill her. Okuda seduces Iulana, and later arranges to have both her and his son killed in a train accident, as alluded to in the novel's title. The son unexpectedly survives the accident, which he obsessively describes, word for word, at several points in the text (17, 54, 106, 141).[38] He begins with a description of the moments leading up to the accident, describing a billboard on top of a building where a children's ballet class is in session as well as the sights, smells, and sounds that accompanied the accident. As the novel progresses, it becomes ever more clear to Shunsuke, who somehow gains access to his father's surreptitiously gathered data, that the event was planned and engineered by his father and his hidden cadre of conspirators.

Okuda acts as though the people in whose lives he interferes are mere avatars in a video game, but the novel insists on physicality and the body. Readers learn that the revered poet abused his own wife and son both physically and psychologically. The poet now watches large numbers of people from a distance, corrupting some and destroying others, while ordering executions and beatings that are described in painful detail. While the repetition of the train accident and the sexual rivalry between father and son underscore a certain mythic inevitability that precludes change, the presence of beaten and dead bodies calls upon readers to question the prerogatives of this powerful and perverse poet.

The novel lends itself to interpretation using Hayles's idea of the third stage of cybernetic evolution, since pattern recognition and randomness characterize cybernetic fictions. The repeated textual passages (patterns) leading up to the accident suggest the mayhem caused by the accident (chaos). Viewed in conjunction with the novel's Oedipal crisis between father and son, the text almost begs for a Lacanian interpretation of desire. For Lacan, the crisis of (male) development corresponds to the idea that desire (the phallus) and subjectivity are founded on absence or lack, which corresponds to a symbolic castration (Hook 62–63). Hayles compares Lacanian desire to the alternation of presence and absence of data as pattern recognition punctuated by randomness similar to the zero-one principle of algorithms used for computer coding (30–31). Desire is always elusive and displaced, according to Hayles, privileging data and absence over embodiment and presence (285). In Cuenca's novel, data, memory, and desire are leveraged or mediated in the images and data retrieved by the son, who often refers to his father as a *lagosta* (lobster). As a metaphorical lobster, the father's phallus (camera) represents the pattern of desire, while his pincers represent castration and absence. The poet refers to his son as *estorvinho* (turbulence), which fits the idea of pattern recognition (the repetition of the exact words of the accident) punctuated by the randomness of the accident (chaos/turbulence). Hayles writes, "Mutation is the catastrophe in the pattern/randomness dialectic analogous to castration in the presence/absence dialectic" (33), and I suggest that perhaps the mutation in this story is an act that may be carried out by the android Yoshiko, which I will describe below.

The train wreck is the central organizing event of the novel, signaling the ultimate absence, death. Although critics have not recognized the novel as science fiction per se, it uses the language of the third cybernetic phase of randomness and pattern recognition, which it deploys along with the presence of the female posthuman, Yoshiko. When Iulana is killed and many bodies are destroyed in the accident, all that remains is the digital archive of the accident data, an archive that can be played over and over. This simultaneously memorializes and erases human existence in a spectral trace. The text also suggests that the posthuman Yoshiko will soon become a substitute for the real body of Iulana in the pattern of desire of father and son. In the final line of the novel, the poet informs his son that he will soon meet Yoshiko when she prepares their evening meal of puffer fish.

Ironically, Yoshiko is the most sympathetic of all the characters, reinforcing Hayles's idea of reflexivity between humans and cyborgs. Yoshiko attempts to understand her own existence as well as the human concepts of death and love, at one point saying that that she wishes she could take refuge in the body of the poet (74), just as he had housed the ashes of his late wife in her android body, as a way of expressing love. However, when Yoshiko spies on the poet with his son's lover, Iulana, she feels a sensation of heat in her chest and a desire to kill him. We realize that Yoshiko, the artificial human, has developed human emotions and may kill in a moment of passion. Since the poison of the puffer fish that Yoshiko is to prepare for dinner can be fatal, we are left to wonder whether she will follow her programming and prepare the meal correctly, or commit a random act of subversion and vengeance by poisoning her master, thereby freeing herself and reinforcing the title of the novel and its "happy ending" (*O único final feliz*). My guess is that, like Paolo Bacigalupi's heroine in *The Windup Girl* (2009), Yoshiko will indeed overcome her programming and pursue her independence. The fact that Yoshiko's inner monologues appear in red ink in the novel, suggesting the color of blood, make this betrayal or transition seem possible. It also may mark what Hayles notes as the moment of rupture of a mutation in the pattern / randomness dialectic: "Flickering signification brings together language with psychodynamics based on the symbolic moment when the human confronts the posthuman" (33), or, as in Yoshiko's act of resistance, the posthuman confronts the human. Yoshiko, ironically, may be capable of the disobedience that the son could not muster, making us question the human / machine divide and the origins of rebellious acts as survival characteristics of the baroque ethos.

Particularly noteworthy in this novel is the Japanese setting. While one critic, Marcel Vejmelka, views Cuenca's use of Japanese culture and setting as Orientalist and superficial (225), it does serve the function of distancing readers from Brazil and its realist paradigms. As a distancing device, this mimics Suvin's concept of cognitive estrangement. The Japanese setting provides an original lens for viewing such issues as illegal immigration, technology, violence against women, data gathering, and generational rivalries, all of which are themes germane to Brazil. It also calls attention to the fetish of foreign culture in Brazil and the receptivity to outsiders, especially those perceived as having technological, financial, or cultural prestige.

Ironically, Cuenca takes aim at the cult of the artist/poet, attacking the fetishism and reverence for high art in Brazil, an approach similar to Rubem Fonseca's subversion of the crime genre in *A grande arte* (1983). In this best-selling novel, which took the literary establishment by storm, Fonseca uses classical culture and literary allusions—features of "high art"—to offer a scathing critique of Brazilian society and its literary establishment while also bridging the popular and the erudite (Vieira 109–12). Cuenca's combination of "fetishizing" high art by "mixing in" science fiction is therefore also part of the novel's anthropophagic devouring of codes or literary *codigofagia* in its resistance to literature as capital.

Conclusion

What is fascinating about the appearance of female cyborgs in different moments of economic and technological change in Mexico and Brazil is how they deploy the gendered body as feminized avatars, both as laborers—as potential brides or sexual partners, engineers or performers, assassins or sex workers—and as mourning humans, struggling for survival through resistance. In their struggle to reconfigure conventional notions of family or love, they remain vulnerable as they attempt to break traumatic cycles and mourn the past. Indeed, all three periods of cyborgs are associated with mourning. In the late nineteenth and early twentieth centuries, the skull-like or zombie state of proto-cyborgs reminds us of slavery and racially marked bodies excluded from modernity. At mid-century, Arreola's and Queiroz's sexual cyborgs dismantle myths of freedom and contrast with the motif of death in gothic stories of violence and exclusion. At the end of the economic miracles of the 1970s, Abreu's Robhéa has a monument constructed in her honor, while Glinda's Penumbria is an elegy of mourning for the failure of the state. By the twenty-first century, Bef's detective Crajales ironically mourns his dead friend Salgado, whom he murdered, while Shiroma retains a connection to her deceased mother, and Yoshiko's body houses the funeral urn containing the ashes of the poet's abused spouse. Their artificial bodies are also reminders of the fallen: the slaves or indigenous masses in

Capitão Mendonça or *Querens*, the economically excluded in tales by Arreola and Queiroz, or the disappeared in "Robhéa" or "Rudisbroeck." The demythification of *mestizaje/mestiçagem* and inclusion in the twenty-first century is exemplified by the victims of crime and urban violence, human and organ trafficking, and organized crime in *Gel azul, Shiroma*, and *Final feliz*.

Each cyborg in these stories is an iteration of the *ethos barroco* of survival and resistance, tradition and innovation, in uncanny layerings and couplings of the human body and technology. The cyborg interface between human and machine does not ultimately lead to a facile sense of mixture of *mestizaje* or new beginnings, but rather to negotiations of citizenship and identity in a complex world of baroque posthuman resistance. In their *Posthumanism and the Graphic Novel in Latin America* (2017), Ed King and Joanna Page examine how the posthuman baroque can unsettle binaries and produce new mutations or outcomes (90), in a way similar to Yoshiko's role in disrupting the system in Cuenca's novel or Shiroma's success in escaping her captors.

In a movement toward more sophisticated interfaces between human and machine, these female cyborgs illustrate issues surrounding embodiment and new forms of the body politic that attempt to transform society into a posthuman baroque or *codigofagia* freed from the hierarchy of the organic and inorganic, offering new configurations of the Latin American posthuman that pay homage to the past while recognizing untapped potential for transforming the future. By the twenty-first century, neoliberal policies and practices are well established, and privatization, crime, and fluctuating markets signal the new digital rhythm of the global economy. However, the gendered cyborgs from this period contest the power of disembodied data or spectral wealth—the monetary fortune of the Cubil family (in the case of *Gel azul*), the economic power deployed by organized crime (in the Shiroma stories) or the destructive potential of digital data recordings (in *O único final feliz*). The networks of capital, crime, and data illustrate the embedding of the virtual world and physical violence against which each of the cyborgs struggle in her own way, resisting bondage or fetishism. While there is a general movement toward more sophisticated interfaces of human and machine, we can see that Echeverría's *codigofagia* can be applied to the Latin American posthuman, who exists between the organic and inorganic and uses embodiment for resistance, giving us new ways of imagining the body politic.

CHAPTER 2

The Baroque Ethos, *Antropofagia*, and Queer Sexualities

Just as the baroque ethos can be used to explain techno-fictions in which cyborg bodies effectively resist economic oppression and modernization, as shown in Chapter 1, it is equally possible, in my opinion, to apply this theory to stories that emphasize sexuality. I base this interpretation in part on Judith Halberstam's *The Queer Art of Failure* (2011), in which she examines queerness, desire, passivity, anarchy, and sexuality as challenges to conventional concepts of class and power. Through her analyses of gender and nonconformity in animated films, photography, and works of art, she finds that the queer—which I would argue in this context largely functions as the baroque—can "sort through the contradictions of capitalism to illuminate the oppressive forms of governance that have infiltrated everyday life" (17). Many of the noncanonical works of science fiction and fantasy that I will examine in this chapter use queer sexualities to challenge conventional heroic masculinist narratives of struggle and nation building. These texts from Mexico and Brazil offer alternative models to the heteronormative unions that underlie the foundational myths of *mestizaje/mestiçagem* and national identity as described by Sommer in *Foundational Fictions* (1991).

In his 2015 study "The Anthropophagic Queer," João Nemi Neto analyzes several contemporary realist works by Brazilian writers and filmmakers to show that the typical ways of expressing queer sexuality in Brazil—and possibly other Latin American countries—are less confrontational than the North American paradigm of "coming out."[1] Neto recognizes the power of LBGTQ movements in the US but emphasizes that Oswald de Andrade's *antropofagia* and its critical assimilation and reformulation of foreign models is essential to an understanding of similar movements in Brazil, where subtlety and indirect strategies are more likely than open protests to bring about a desired result.

Echeverría's thesis is that, during the earlier period of colonization, strategies for survival resulted in a cultural *codigofagia* that reformulated social and cultural hierarchies. These reformulations are apparent in the baroque art of sixteenth- and seventeenth-century Latin America, where local artisans reworked Catholic iconography to incorporate local or popular deities into ecclesiastical art, challenging both the church and the precepts of art history and its exemplary artists. Echeverría has noted that both the indigenous and Afro-descendants ended up transforming the baroque, revitalizing an art form and creating a distinct imaginary, "una Europa latino-americana" (*La modernidad barroca* 82; a Latin American Europe).

Brazilian poet and essayist Haroldo de Campos emphasizes that what Echeverría characterizes as a revitalizing gesture was not mere passive imitation, but an anthropophagic reinterpretation of European culture. De Campos cites the baroque (in art and poetry) as examples of the resistance and omission that contest the linear historical narrative of Brazil's colonization: "Já no Barroco se nutre uma possível 'razão antropofágica,' desconstrutora do logocentrismo que herdamos do Ocidente. . . . É uma antitradição que passa pelos vãos da historiografia tradicional, que filtra por suas brechas, que envieza por suas fissuras" (17; Already in the Baroque a possible 'anthropophagic reason' arises, which destroys the logocentrism that we inherited from the West. . . . It is an anti-tradition that passes through the gaps of traditional historiography, which filters through its gaps, skewing it through its fissures.) De Campos portrays the baroque as an antecedent to *antropofagia*, as an "anti-tradition" that resists and deconstructs the discourse of nation building while emphasizing distortions and omissions in official histories.

I maintain that the "anthropophagic" or "queer" baroque—which emphasizes fissures, absence, passivity, or scarcity—is distinct from the well-known esthetic of the "neo-baroque" in Latin American literature, which in the writings of authors such as Severo Sarduy, Lezama Lima and Néstor Perlongher emphasizes excess, energy, proliferation, and profusion. According to Sarduy, for example, the basic tenets of neo-baroque esthetics establish relationships among artistic creation, writing, sexuality, and, more particularly, homosexuality. In his view, the baroque can be illustrated in terms of the energy expended in non-procreative sex: "How much lost work! What a lot of play and waste! What a lot of lost effort! . . . Play, loss, squandering and pleasure, eroticism as an activity which is always purely playful, which is no more than a parody of the function of reproduction, a transgression of the useful, of the 'natural' dialogue of bodies" ("The Baroque" 130). In general, Sarduy sees the ornamental nature of the neo-baroque as a critique of the biopolitical logic of sexual and economic productivity in heteronormative society ("La simulación" 99–100). Similarly, Argentine writer Néstor Perlongher, who resided in Brazil for many years, promotes the idea by writing in the esthetic of the "homobarroso" in the 1990s, in which he establishes a connection between homosexuality and the neo-baroque style of parody, carnivalization and camp (115–16), again emphasizing an esthetic of intensification and excess.

This interpretation of the neo-baroque contrasts with ideas set forth by Silviano Santiago in his article "O homossexual astucioso" (The wily homosexual) in which he argues that queerness can and should be expressed subtlety. Santiago surveys the tactics of North American gay movements and wonders if a slower process would not lead to more dialogue and long-term social cohesion, asking:

Se formas mais sutis de militância não são mais rentáveis do que as formas agressivas? Se a subversão do anonimato corajoso das subjetividades em jogo, processo mais lento da conscientização, não adiciona melhor ao futuro diálogo entre heterossexuais e homossexuais do que o afrontamento aberto por um grupo que se auto-marginaliza, processo dado pela cultura norte-americana como mais rápido e eficiente? ("Astucioso" 15–16)

What if more subtle forms of militancy could be more effective than the aggressive forms? What if the subversion of the courageous anonymity of subjectivities in play, a process slower than consciousness-raising, could contribute more to a future dialogue between heterosexuals and homosexuals than open confrontation by a group that self-marginalizes, the process considered by North-American culture to be the fastest and most efficient?

Denilson Lopes's 2008 article "Silviano Santiago: Estudos culturais e estudos LGBTs no Brasil" offers a survey of Santiago's work in comparative perspective with North American gay movements. He emphasizes that Santiago's commitment to dialogue is not simply avoidance of responsibility or an endorsement of Brazilian "openness" to sexuality, but instead is a pathway toward building solidarity with other marginalized groups who seek to move beyond individual resentment and anguish (951). The social body and social solidarity are key concepts in Santiago's *conscientização* in Brazilian LGBTQ movements, as substitutes for the American concepts of self-expression, individual rights, and identity politics.

Santiago's idea of *malandragem*, or the "wiliness" of Brazilian gays, is an extension of his idea of the "space in-between" that emerged from Brazil's anthropophagic tradition. This places writers between rebellion and clandestinity, in a contingent space in-between the national and the international cultural scene ("O entre-lugar" 15–16).[2] Nemi Neto affirms that Santiago's liminal environment allows a different way of expressing the queer to take shape in Brazilian society, but describes his own approach as anthropophagic, in that it combines elements from both American and Brazilian cultures (5).[3] While Nemi Neto primarily examines works dealing with male homosexuality,[4] I will examine queerness in a more general sense. Some of the stories I will analyze have characters who exchange male and female bodies;

in others, we find societies that are all female, or that practice cross-
dressing or even share bodies. We even find female characters who
fashion themselves as other life forms, including amphibians and
marine life. Like sexual minorities within society, such literary characters inhabit a liminal or queer space, contesting normative social
arrangements through gender experimentation.

I would suggest that Echeverría's baroque ethos is similar to Santiago's space in-between as a strategy of resistance, but differs from
it in assigning a more central role to a sense of belonging. This is
brought up by Gazi Islam in his analysis of the relevance of antropofagia for Brazilian culture and its sense of embodiment, because isolation does not fulfill the social need for acceptance by others, since
"the body's predicament [is] being both separated from other bodies,
yet needing nourishment form the outside" (172). This "devouring"
of the other can be culturally affirming as an "appropriation of the
other within the self" (172). In this way, the other is not pushed aside
but engaged in an "astute" or indirect way.

Before embarking on the analysis of Mexican and Brazilian texts, I
should note that, beginning in the seventies, Anglo-American science
fiction and fantasy begins to treat the topics of sexuality and gender.
Speculative fiction offers possibilities not only for new landscapes but
also for new beings, many of which may contest conventional concepts
of gender and identity. In her study of women and fan magazines of
the Golden Age, *The Battle of the Sexes in Science Fiction* (2002), Justine
Larbalestier explains how science fiction has challenged conventional
concepts of gender: "In these texts there are often many more sexes
than two or even three. Being a man or a woman is not a given; it
is always unstable" (13). Brian Atteberry's *Decoding Gender in Science
Fiction* (2002) also studies women in science fiction, analyzing female
characters and authors of the genre from its beginnings during the
Golden Age to the feminist revolution of the 1970s and beyond. As
Rob Latham points out, even when science fiction claims to shy away
from sexuality, it is "always treating sexual topics, perhaps most pow-
erfully when it seems to be primly avoiding them" ("Sextrapolation"
53). American texts in which protagonists change sexes include Ursula
K. Le Guin's *The Left Hand of Darkness* (1969) and John Varley's stories

"Picnic on Nearside" (1974), and "Options" (1979). As we shall see, some of the most interesting stories about shifting sexual identity written in Mexico and Brazil appear during the late nineteenth and early twentieth centuries, much earlier than their American counterparts.

I wish to emphasize that my goal is to analyze the portrayal of sexuality and feminism in selected Mexican and Brazilian speculative fiction. A more complete listing of relevant stories is in the notes for both Mexico and Brazil.[5] Historical accounts of women authors and feminism can also be found in several studies, including Gabriel Trujillo Muñoz's chapter on Mexican women writers in his 1999 study *Biografías del futuro* (329–43), and the section on women as characters and authors (205–9) in my 2004 *Brazilian Science Fiction*. In my article "Recent Brazilian Science Fiction and Fantasy Written by Women" (2007), I analyze examples of shifting female voices from three distinct narrative perspectives, while Gabriela Damián Miravete traces a brief history of Mexican women writers of science fiction in her article, "La mano izquierda de la ciencia ficción mexicana" (The left hand of Mexican science fiction), whose title pays homage to Le Guin's famous novel, *The Left Hand of Darkness* (1969). Miravete's article outlines the history of women writers in the genre, and mentions more contemporary women authors who make gender the main focus of their work.[6]

While issues of gender, family, and reproduction play out primarily in the context of extraterrestrials and technology in Anglo-American science fiction,[7] most Latin American authors address these issues by delving into the minds and bodies of fellow humans in tales of embodiment and queer sexuality. In *Transvestism, Masculinity and Latin American Literature* (2002), Ben Sifuentes-Jáuregui states that, although he recognizes the importance of Judith Butler's and Eve Kosofsky Sedgwick's theories of gender performativity, he finds the concept of embodiment to be far more important in Latin American literature:

> In Latin American writing, a body is always sought after and claimed as one's own: this act of possession and incorporation is quite powerfully displayed in literary texts. . . . Although I understand that these bodily transformations are of different registers, what is again important to underscore is that the very physicality of the body be-

comes the *materia prima* for sexual and gender representation. Prior to a philosophical engagement with the body, there is an attachment and possession between the body and subjectivity. (194n7)

While Sifuentes-Jáuregui focuses on male transvestism and the shift from "masculine" to "feminine" identities (4) in literary and cultural studies, I am interested in more complex and multi-directional shifts between genders in the form of distortions and transformations of the body that are not always possible in conventional literature.

In this chapter, I examine embodiment and the baroque ethos in landscapes of Mexican and Brazilian speculative fiction that range from Siam to the remote Amazon to seventeenth-century Mexico, all places where women warriors undermine gender and social structures of colonial narratives. I then move on to stories situated in the belle époque societies of Rio and Mexico City in which women inhabit men's bodies. In the final section I analyze texts in which women are identified or identify with other species, such as amphibians, reptiles, and creatures of the ocean depths. I demonstrate that the characters' passivity and failure or even refusal to participate in conventional behaviors, especially as regards reproduction, reinforce the idea of a distinct ethos of the queer baroque. By analyzing texts from key moments when women and men challenge social expectations and cultural paradigms, I show how subversion through the strategies of the queer baroque recasts protest against traditional social institutions and power structures.

Women Warriors

An early queer text in Brazilian literature, Machado's story "As academias de Sião" (1884; The academies of Siam) takes place in Siam (present day Thailand), a setting displaced in space and time from the author's typical scenarios of nineteenth-century Brazilian society. The story is told mostly from the point of view of Kinnara, a concubine who becomes first a woman warrior and then a king. The story forms part of Machado's collection *Histórias sem data* (1884; Timeless tales), whose title alludes to the fantastic content or settings of its texts, several of which take place outside of historical time and outside of Brazil. Among these stories, both "As academias de Sião" and "Singular ocorrência" (A singular occurrence) deal with female gender identity and sexuality. While Machado generally portrays female sexuality and adulterous females sympathetically, as either victims or willing sexual partners, he is supremely aware of society's contradictions and the double standard for men and women. As the last story in the collection, "The Academies of Siam" is perhaps most radical in its views of gender and society.

Braulio Tavares succinctly summarizes Machado's story, stating that it "shows an effeminate king and his dominating concubine exchanging bodies by magical means, to prove the theory that souls are as gendered as bodies" ("Stories" 86–87). However, Tavares's straightforward reading of the fantastic tale omits a major subplot, in which the debate on the existence of an "ungendered soul" unleashes violence within the society's learned academic community. This recalls Derrida's argument, made in his essay, "The Law of Genre," that the concept of "law" lies at the heart of differentiation and order, where it represses the madness and disorder that lurk underneath (77). Thus, while the law of genre (like gendered souls) purports to shed light on systems that govern society and classify individuals, it blinds us to alternate ways of being (such as the ungendered soul).

This repression is shown in Machado's story in events other than those directly concerning the king with the female soul and concubine with the masculine one. When the four original Academies of Siam first debate the existence of gendered versus ungendered souls, the argument degenerates into physical violence as the proponents of the gendered soul theory secretly arm themselves and kill thirty-eight members from the other three rival academies. When the city awakens to discover the horror of this mass murder, the only person who excuses the massacre is the concubine Kinnara, an apparently "masculine-minded" woman. She manages to get the king to sign a decree stating that the gendered soul theory is the only acceptable and legitimate doctrine, while the other notion is to be deemed "absurda e perversa" (469; absurd and perverse). This pleases the victorious faction and also re-establishes peace and order in the city.

Machado illustrates that binaries and hierarchy are a type of violence that lies hidden at the center of the social order, since the repression of the idea of androgynous or genderless souls exposes cultural concepts of gender as fictions. Whereas nowadays there is a wider spectrum of socially acceptable professions and behaviors for men and women, the society portrayed by Machado (and perhaps his own) had rigidly defined male and female roles to which bodies and souls had to conform. Derrida shows that this strict division illustrates the workings of binary logic in ideology, according to which there must be strict boundaries between categories: reason and madness, men and women, spirit and matter, presence and representation, essence and simulacra. For Derrida, the human desire for a center or "authorizing pressure" orders these hierarchical oppositions. In the story, this authority or "logos" is clearly symbolized by the gendered faction, which maintains the dogma of gendered souls. I argue that this is the logocentrism that the baroque queer deconstructs.

Kinnara talks the king into undergoing a magical ritual that allows their souls to switch bodies for a period of six months. Afterward, the new macho king (Kinnara's masculine soul now in a male body) brings on an era of unprecedented prosperity, as she makes sure that taxes are collected, foreign missionaries are massacred, and, after a brief but glorious war, honor and unity are brought to the land. At the root of each of these activities is violence and death associated with the doubly patriarchal unity of body and mind in the warrior king, Kinnara.

To perpetuate "his" powerful new kingdom, king Kinnara decides to have the concubine killed, but is plunged into a quandary when he discovers that she is pregnant. Ironically, this hesitation begins the queering moment of passivity or space in-between that ultimately undermines all previous patriotic achievements. Instead of killing his pregnant concubine, he asks several of the Academicians for advice, but in interview after interview, he fails to get a clear answer out of any of them, as each claims that the others are idiots. When the end of their six-month agreement arrives, the king is forced to return to his/her original, now pregnant female body, and after having initially felt the elation of paternity, Kinnara now experiences the physical and emotional sensation of maternity. The text does not speculate on what it was like for the King Kinnara to impregnate her former female body or, for that matter, what it would be like to kill it.

The story ends with pregnant Kinnara wondering why the members of the academy are idiots as individuals but enlightened as a whole. This speculation is an allusion to the consensus or shared notions that allow society to function. By exploring arbitrariness of gender and the violence of repressing queerness, Machado touches on Derrida's dictum that "one cannot conceive of the truth without the madness of the law," (76), where *law* refers to the fictions of difference that sustain society and the organization it imposes on gender. In the end, Kinnara seems to conform to societal norms and the physical exigencies of her body. Machado's text thus belongs to a line of philosophically inspired fantastical texts that deal with epistemology and our knowledge of our world as illustrated by our own restricted or conventional notions of gender as well as the violence that surfaces when such dictums are not followed.

While the Academies enforce the law of gender, ironically Kinnara and the king live a queer existence, contesting the law that enforces traditional gender roles and prescribes that women must be passive and men aggressive. Yet the story comes full circle after Kinnara's interlude as a warrior king. She is returned to a space in-between. She knows that beliefs attributed to race, class, and gender, even when proved a fiction, are used to serve power, as Machado was well aware. Thus, pregnant Kinnara, who appears to conform, is fully aware of the arbitrariness of gender identity.

Women warriors are also found in Gastão Cruls's *A Amazônia misteriosa* (1925; The mysterious Amazon), a tale about European contact with the descendants of the original tribe of Amazons from Greek antiquity. In this novel, Cruls imagines a gynocracy of Amazon women, whose values anticipate the concept of an anticapitalist matriarchy as outlined by Oswald de Andrade in his "Manifesto Antropófago" (1928; Cannibalist manifesto), which he further developed as a component of his alternate model of matriarchal techno-modernity in his 1953 "A marcha das utopias" ("A marcha" 205–9).[8] As a work of science fiction, *A Amazônia misteriosa* inhabits the space in-between because, as Roberto Causo points out, it absorbs and critiques H. G. Wells's *The Island of Dr. Moreau* (1899), while evoking local cultural elements: "Gastão Cruls realizou o feito de absorver a influência estrangeira para produzir, a partir dela e de influências locais de caráter mítico, uma obra original" (*Ficção* 177; Gastão Cruls carried out the feat of absorbing foreign influences in order to produce an original work, combining both foreign, local and mythic features).[9] In this quotation, Causo essentially describes the novel as one of the first examples of cultural *antropofagia* in Brazilian science fiction.

As I have argued in "The Amazon in Brazilian Speculative Fiction: Utopia and Trauma" (2015), the Amazon region functions as an "empty signifier," susceptible of being incorporated into any number of myths proposed by Brazilian science fiction authors. Historically, as Candace Slater has noted, the Amazon has traditionally been portrayed in contradictory ways by explorers, either as a remote natural paradise filled with unattainable riches, or as a verdant inferno resistant to the forces of modernization. These views of the "uncanny" or the "monstrous" Amazon blind us to the region's colonial realities and the struggles of its actual inhabitants (38–48). The standard image of modern-day Amazon women warriors begins with early chroniclers who, in documenting their perverse, cannibalistic practices, transform them into an allegory of the untamed Americas (Braham, "Monstrous Caribbean" 24). Slater, however, downplays their cannibalistic practices, emphasizing their status as warriors. She notes that during his 1541–1542 journey down the Amazon, Dominican Friar Gaspar de Carvajal has almost nothing bad to say about the women of the Amazon region, instead praising them for their valor and elegance. Slater notes that Carvajal's "lack of criticism is strange since these women's sexual license, lust for battle and fondness for graven images would appear cause for the strictest censure on the part of a Roman Catholic priest" (93). She notes that, like the Amazon itself, the women warriors resist classification as projections of both fear and desire (90–91).

In *A Amazônia misteriosa*, Cruls quotes Carvajal's account (159) and also treats Amazon women characters with deference and wariness. Predictably, therefore, his novel demonstrates an attitude distinct from that of contemporaries such as Menotti del Picchia and

Jeronymo Monteiro, for whom Amazonian women are part of the "lost civilizations" trope and linked to Viking culture or the myth of Atlantis.[10] Cruls's Amazonian novel, while sharing some features with these Brazilian "lost world narratives," exhibits familiarity with readings about Amazonia by naturalists and explorers, mentioning, in addition to Carvajal, Alexander von Humboldt, Alfred Wallace, Henry Walter Bates, and Brazilian Cândido Rondon, all of whom showed respect for the landscape and the people of the Amazon in a way "lost world" novels do not.

Although Cruls largely invents the culture of the Amazonian women in his novel, he does connect them to indigenous colonial history and conquest. The protagonist of the novel, a Brazilian doctor, gains this perspective after he drinks a hallucinogenic substance that grants him the power to see the past as if by "telepatia e televisão" (106; telepathy and television). He feels as if he is suddenly flying over the great monuments of pre-Columbian civilizations, including Tenochtitlán, the Temple of the Sun, and the cities of Cusco and Quito. He is even able to meet and understand the last emperor of the Incan empire, Atahualpa (1502–1533), from whom he learns about the violence of invasion, rape, and subjugation of his people at the hands of Pizarro and his troops. The narrator learns that the Amazons are likely descendants of Incan women who rebelled after this defeat, abandoning their husbands, making their way to the Amazon to found a new civilization: "Daí o seu gesto de rebeldia, que as levou a trucidarem desapiedosamente todos os filhos varões, e consequente lembrança de formarem uma sociedade exclusivamente composta de mulheres" (85; Hence their gesture of rebellion, which led them to slaughter pitilessly all male offspring, resulting in the memory of forming a society composed exclusively of women). The fact that they abandon their male children provides an important detail for the plot, since the novel centers on a German scientist, Jacob Hartmann, who finds in the Amazon's abandoned male children a steady source of human subjects for his neurological and cross-species breeding experiments.

A Amazônia misteriosa is a hybrid work that combines an appreciation for the history, ethnography, and local color of the region with a science fiction plot. When the first-person narrator, the unnamed Brazilian doctor, and his companion, a simple country man, find themselves separated from their expedition, they are rescued and led to Hartmann's compound by members of a peaceful indigenous tribe. When the Brazilian doctor finds out about Hartmann's cross-fertilization experiments on humans and animals, he accuses the German of being a type of Dr. Moreau. Hartmann counters by saying that his experiments have led to advances in the treatment of aphasia, in human rejuvenation, and in cross-species breeding.[11] Since the scientist does not want his practices to be exposed to official investigation, he informs the doctor that he and his companion must remain as prisoners until he is ready to share the benefits of his experiments with society.

Unlike the desirable exotic princesses of "lost world" narratives, the Amazon women in *A Amazônia misteriosa* are warlike. When the members of the European compound attend the tribe's ritual celebrations, including a hunt, voluntary submission to painful ant bites, and an archery contest designed to identify their future queen, the Brazilian doctor is impressed, an attitude that contrasts with the moral condemnation or fear of female sexuality found in myth and legend. The story also contrasts with the typical heteronormative romance formula of rescue, in which a European male conquers a native princess, since the Brazilian doctor tries to flee from the compound with the scientist's wife, whom he prefers over an Amazonian woman who shows interest in him.[12] However, the story ends tragically when the wife falls behind, is wounded by an arrow, and plunges into the river while trying to escape.

Thus, the narrator's Amazonian adventure is ultimately one of failure, which is part of its anticolonial message. While the protagonist wishes to protect the Amazonian people from the German scientist's human experiments, he fails to do so.[13] Cruls's protagonist also fails in romance, disengaging the novel from the gendered paradigms of conquest and heroic male action typical of this subgenre. The Amazon women, whose history and culture queer the conventions of the lost world adventure novel, also challenge the capitalist, patriarchal concepts of private property, monogamy, work, and competition.

Indeed, the Amazonian women generally remain untouched by Brazil's patriarchal society. Their presence in the region, coupled with the novel's dénouement, typify what Halberstam has called the "queer art of failure," which creates alternate or imaginary spaces where social organization diverges from that of the Oedipal family and the premises of "civilization."[14]

I now turn to another rewriting of the heroic narrative of colonial conquest in Carmen Boullosa's *Duerme* (1994; Sleep), a novel that takes place in sixteenth-century Mexico City and its environs. The female protagonist of the novel, Claire Fleurcy, is loosely based on the historical figure of Catalina de Erauso, a Basque nun who left her convent in Spain, arrived in Mexico in 1603, and fought in battles in Chile and the Andes. Boullosa's novel is often overlooked as a work of the fantastic given its colonial setting, yet queer elements appear throughout the novel in the guise of indigenous medical and religious practices as well as rewritings of legends and fairy tales integral to its message.

As Jill Kuhnheim points out, Catalina de Erauso's use of male attire during this baroque period allowed her to escape abuse and gain status in society, including recognition by the pope as a courageous and virtuous woman (13–14). Stephanie Merrim also notes that cross-dressing during the period of the Spanish baroque was generally accepted, and that the adjective *varonil* (manly) was a positive indicator regardless of gender (188). In *Duerme*, Claire is captured while in pirate attire and transported to Mexico City, where she is forced to dress like the Count de Urquiza, who is scheduled to be hanged for opposing the Crown. She is then submitted to a strange procedure whereby her blood is replaced with sacred waters by Inés, an indigenous woman charged with her care. Thus disguised, she is hanged on the gallows but revives because the sacred waters have made her immortal. However, she learns that if she strays more than seven leagues from Mexico City, she will fall into a deep coma, like Sleeping Beauty—hence the novel's title, *Duerme*. Thus, she has been given the gift of immortality with restrictions of confinement and passivity.

After Inés's intervention, Claire considers herself to be "self-born," declaring "soy mi propio hijo" (19)— illustrating through this masculine pronoun a moment of queering of the text. During the procedure, Claire comments, "Veo cómo una vena, en un gesto excepcional, bebe del agua, como se fuera la garganta sedienta de un polluelo" (20; I see how a vein, as an unusual gesture, drinks of the water, as if it were the thirsty throat of a chick). This image of the throat, a displacement for the birth canal, is associated with a chick, an offspring that births itself as it chips out of its own egg. The imagery is that of a reverse-engineered birth during which Claire is endowed with the power to heal herself. Throughout the novel, her use of clothing emphasizes a lack of gender identity as she performs different genders and races, with no essential self (Taylor 227). When she is in a male role, she fights and uses masculine adjectives to describe herself. Although she is white, when she wears the traditional clothes of indigenous women, she experiences the harsh reality of physical labor and rape at the hands of the Conde de Urquiza, who summarily uses her. Claire's ability to move between worlds and to do what "works" challenges the conventional Western sense of a stable identity, thus illustrating the baroque ethos in its accumulation of experiences, resistance, and survival.

Claire's gender destabilization is also paralleled by the unstable mix of genres that compose the text, which include a baroque play about a love triangle of Aphrodite, Ares, and Hephaestus, songs and verse, fairy tales, indigenous pictograms, and historical chronicles. Although Claire never literally changes into a man, when she goes into battle to put down a rebellion in Querétaro, her sexual identity is inadvertently revealed after she is wounded. She is returned to Mexico City where her companion Pedro de Ocejo cares for her, along with an Italian actress, Afrodita, with whom she shares the joys of love. The actress refers to Claire as her Ifis (Iphis), a mythological figure who changed sex from female to male. According to Deborah Kamen, the story of Iphis is the only "mythological account of same-sex desire, not only in Ovid but in all of classical literature" (21). In Ovid's *Metamorphoses*, Iphis, although born a girl, was raised as a boy in order to protect her from her father's prejudice. After falling in love with a young woman, Iphis prays to Juno to be changed into a man in order to marry, a request to which Juno accedes. The queer overtones are not lost on Boullosa, who inserts Iphis into a baroque allegorical play that Afrodita and Pedro de Ocejo write for Claire's entertainment. Several critics have used Butler's theories of gender performativity in analyzing Claire's shifting use of gender, yet few have examined the implications of the magical female-to-male change of this classical personage in the novel, nor, until recently, the role of theater and theatricality in gender performativity. In her 2018 article, María Cristina Pons uses Majorie Garber's idea of the "third actor," used in classical works of Greek theater, as a way of destabilizing binaries. In the case of *Duerme*, Pons asserts that, as the third actor, Claire enacts the "efecto travesti" (transvestite effect) that allows her to break down categories of sex, gender, and race, on the one hand, and human and immortal, on the other (65–66). Notably, Pedro Ocejo walks in on the love scene between Claire and Afrodita, who again requests that he include Iphis in his play. He refuses and takes her into his own bed, effectively reasserting heteronormative roles.

Soon after, Pedro de Ocejo has to remove Claire from the city where she is wanted by the authorities, with the result that she enters a coma, as ordained by her treatment with sacred waters. Pedro narrates the end of her story twenty-five years later, as if to give it closure, saying that he failed to revive her. Yet in the final chapter, Claire reappears in the narrative, telling how she was revived and revealing her plans to lead a rebellion against the Spanish and build Mexico. As Pirott-Quintero notes, Claire ultimately controls and re-opens the narrative, since "her 'story' is not finished . . . but is reactivated and continued each time the reader 'awakens' her with her reading" (7). Thus, Claire is Sleeping Beauty, whose utopian rewriting of history is procreative in the literary sense, activating and disseminating her narrative with each new reader.

Claire's story in *Duerme* combines history and narrative from the point of view of those outside of official history and colonial experience, feminizing traditional chronicles of male-focused battles in a tale of queered cultural history. Claire's self-generation frees her from the stigma of female betrayal and guilt of womanhood in Mexican culture. As Pirott-Quintero points out, Claire foregoes any type of traditional reproductive role, breaking with the notion of blood associated with *mestizaje* and the condemnation of La Malinche (3), a figure traditionally associated with the betrayal of her people and the iconic mother of Mexico's mestizo population.[15] Claire is a queer icon of Mexican identity: while she cannot vanquish and conquer in a traditional sense, she can draw supporters together at the center of Mexico, navigating both Hispanic and indigenous cultures. In her space in-between, Claire exemplifies the queer baroque ethos because she chooses gender performativity over essence, sleeping over consciousness, and passivity over action.

All of these women warriors can be conceptualized as thought experiments about gender, power, and nationhood. Machado's Kinnara makes us question the naturalness of gendered behavior and motherhood, while Cruls's Amazonian women and failed expedition counter heteronormative colonialist paradigms. As Jill Kuhnheim explains, in the case of Boullosa's *Duerme*, the novella is not so much about a character, but "the questioning of the subject, of master narratives, conventional genres and representation itself" (9) that lies at the center of nation building. Women are given the opportunity to participate in positions of power in war, sexuality, and even politics, yet choose to remain a part of a body: either a social body, through procreation, in case of Kinnara, or a cultural body, as in the case of the Amazonian tribe, or a literary body, as in the fairy tale of Claire. Although characterized as women warriors, none participate in the conventional narrative of conquest, queering expectations while offering alternatives of community and survival characteristic of the baroque ethos.

Women in Men's Bodies

Women warriors, although resistant to heteronormative culture, reproduce either physically—by participating in procreation—or culturally—by generating national narratives. In the case of texts about women in men's bodies, we see a break from participation in conventional reproduction and nation building. As women begin to share spaces typically reserved for men, narratives turn from nationhood and power to issues of trauma and embodiment. When women invade traditionally male realms, social and sexual conflicts arise, including fear of alternate sexualities that erode traditional narratives of nation and masculinity. In her 1985 study of nineteenth-century English literature, *Between Men*, Eve Kosofsky Sedgwick describes clubs, social activities, and work as "homosocial" spaces, where men relate to one another and women are viewed as objects of exchange. As such, Sedgwick has defined this area as "a problematical *space*" that questions "the whole package of physical and cultural distinctions between women and men" (29).

Since there are no established laws against homosexuality in Mexico or Brazil, the fear of "sexual anarchy" or "homosexual panic" that occurred with the trial of Oscar Wilde was not as great in these countries as it was in England. As Silviano Santiago notes, the open expression of queerness was already present in Brazilian naturalist works of the 1890s ("O homossexual" 13–14), several of which are now classics read as part of secondary education in Brazil.

The late nineteenth- and early twentieth-century period was marked by a growing rift between science and the arts, when anxiety about evolution and the scientific method led writers to delve into the occult. Mark De Cicco has used the polyvalent word *queer* to examine the role of the occult sciences in literature during the Victorian period, when the term meant peculiar (or strange), eerie (or supernatural), or transgressive (either physically or sexually) (5). For De Cicco, the trajectory of the scientist/protagonist during this period who delved into the occult was moving toward a space of the queer. The fascination with Theosophy, magnetism, and other forms of the occult was also common in Mexico and Brazil (Haywood Ferreira 168), but because religious syncretism and embodied practices of popular Catholicism was already established here, the sense of entering forbidden territory is less pronounced, as if these cultures were "inoculated" or prepared for their contact with the queer sciences.[16] In his analysis of Robert Louis Stevenson's *The Strange Case of Dr. Jekyll and Mr. Hyde*, De Cicco notes that "the attempt to come to terms with the queer or abnormal, as well as the scientist's survival, depends upon the ability of the scientist to 'queer' himself and/or his subject in preparation for the encounter with the occult"

(7). This idea is useful for the interpretation of Mexican and Brazilian narratives where women enter men's bodies.

An early text that speculates about a man sharing his mind and body with a female spirit is Mexican Amado Nervo's *El donador de almas* (1899). Oddly enough, Nervo's text is set in Mexico City in 1886, the year of the publication of Stevenson's *The Strange Case of Dr. Jekyll and Mr. Hyde*. Nervo's text is told from the perspective of physician Rafael Antigas, a lonely man of science. Pining for a female soul mate, Rafael has an encounter with the queer or esoteric science when his friend Andrés Esteves uses his occult powers to provide one for him. When Andrés releases the spirit of the female Alda from the body of a comatose nun, her spirit accompanies the doctor, helping him build his career and improve his bedside manner while offering companionship. She confesses, however, she cannot offer him love, since her soul is controlled by Andrés. When the body of the nun dies, Alda's soul urgently needs a new body or she will be lost to Rafael forever. Eventually Andrés chooses the left side of Rafael's brain for this purpose. She enters, regains her own free will, and discovers that she does indeed love him. At this moment Rafael experiences a "beso mental" (36; mental kiss), a transference of courtship conventions to the spiritual realm.

Although critics have generally focused on *El donador de almas* as a struggle between two types of knowledge—the scientific and the occult—I find it has much to say about gender, the body, and embodiment.[17] When Alda becomes part of Rafael and they experience their first moment of passion, the physical aspect of this encounter does not go unmentioned, making Rafael blush.[18] By caressing himself, Rafael acknowledges his sexual feelings and physical urges, convincing himself that Alda understands the weakness of the flesh. Thus, Alda experiences a queer moment of sexuality as she perceives his male manifestations of physical desire.

The queer is also apparent in the fact that Alda acts as an extension or double of Andrés, since both share interests in literature, poetry, and the occult, which Rafael does not. Analyzing their relationship as a type of love triangle, we see that Rafael and Andrés form what Sedgwick has called a homosocial bond, while Alda remains an outsider, a

mere object of the men's relationship. This is especially evident after
their initial honeymoon phase, when Alda and Rafael's love quickly
ebbs, and Rafael starts to project the irritation he feels toward Alda
onto the poet Andrés.[19] As De Cicco points out in his discussion of
Victorian gothic texts, the doctor or scientist in such stories desperately
tries to separate himself or "straighten" the queer desire he experiences. Subsequently, Rafael feels an urge to rid himself of Alda, "ese
cuerpo extraño" (57; that strange body) that inhabits him. Eager to
leave Rafael's body, Alda complains of being caught in Andrés's liq-
uid "magnetism" or viscous fluids used to tie souls to bodies, which
I interpret as an intermediate state between male and female (Grosz
203). This experience of maleness is what Alda eventually comes to
experience as imprisonment.[20] Later, after Andrés performs an incan-
tation to free Rafael from Alda, proposing to place her spirit into the
body of his housekeeper, Alda chooses to live without a body in a
completely spiritual realm shared with others like herself.

Another queer moment occurs when Alda describes herself as
having male gender: once she takes up residence in Rafael's brain,
she exclaims, "me siento hallado" (35, emphasis added; I sense that
I am found). Alda and Andrés queer the doctor Rafael, who ends
up writing a poem, "Tenue" (Tenuous), which closes the story. The
poem contains synesthetic or sensory combinations that suggest his
unstable and contradictory states. These include "fantasma de un
eco" (ghost of an echo), "sombra de un suspiro" (shadow of a sigh),
and "alma de perfume" (76; soul of perfume). The first combines the
supernatural with the aural, the second the visual with the physical,
and the third, the spiritual with the olfactory. While this marks the
style and mystical experience typical of *modernismo*, each suggests a
queer way of thinking beyond binaries, using synesthesia to reinforce
the queer themes of the story.

The story of Rafael, Alda, and Andrés questions the binaries of
body and mind, life and death, female and male, slave and master,
art and science. Alda first experiences male sexuality, then escapes
her male masters (Andrés and Rafael), choosing to "live" outside or
beyond the codes of society. Rosemary Jackson observes that, in many
texts of the fantastic, categories break down and everything appears

to operate according to a principle of correspondences, resulting in a "collapse of differences" (50), as when the three main characters of *El donador de almas* begin to function as doubles of one another. She further notes that, "Doubles, or multiple selves, are manifestations of this principle: the idea of multiplicity is no longer a metaphor, but is literally realized, [and] self transforms into selves" (50). Here the stability of identity is questioned, just as in the poem "Tenue" itself. In Jackson's view, such doublings are mainly "concerned with erasing rigid demarcations of gender and genre" (49).

Ultimately, the tenuous relationship cannot be sustained among the three. Alda's strategy is thus one of evasion, a baroque resistance toward conventional gender roles. Rafael recognizes the failure of his experiment with heterosexual love, although he gains an appreciation for the arts. Although their pairing fails them both, as Judith Halberstam notes, "gender failure means being relieved of pressure to measure up to patriarchal ideals" (*The Queer Art* 4). In the mock interview at the end of the novella *Donador de almas*, the fictional character/"author" articulates his own baroque ethos of resistance through his own "lack of seriousness" in writing about gender instead of Spain's loss of its last colonies in the 1898 Spanish American War. He calls his text a *nouvelle*, not only because it is brief, but also because it represents generic mixing, with elements of humor, poetry, literary interviews, visions of utopian worlds, horror, and the esoteric. By apologizing for writing about sexuality, Nervo is admitting to failure in Halberstam's sense of being a subversive intellectual, since he refuses to "accede to the masculinist myth of Herculean capitalist heroes who mastered the feminine hydra of unruly anarchy" (*The Queer Art* 18), articulating instead the quiet queer art of failure that questions the heteronormative narrative of colonization, conquest, and nation building. This comment seems particularly appropriate for the historic moment when Spain loses its last colonies.

Unlike Nervo's oeuvre, which has remained of interest to scholars, the works of Brazil's Coelho Neto have generally fallen out of favor. One of these, however, the novella *Esfinge* (1908; Sphinx), rediscovered by Roberto de Sousa Causo in his 2003 study on the history of speculative fiction (*Ficção* 112–16), is notable for its queer narrative. This novella, like *Donador de almas*, is also characterized by its queer protagonist and a fascination with the occult sciences. However, unlike Nervo, Coelho Neto explores the potentially unsettling effects that a queer person can have on others, using the setting of a semiprivate homosocial space of a boarding house in Rio de Janeiro. The residents of this house are mostly middle-class, single white men; the only two female boarders, both English, are the motherly Miss Barkley, who runs the boarding house, and Miss Fanny, a young, red-haired teacher. One other resident, the Englishman James Marian, is a tall, well-formed man who has an uncannily feminine face.

Coelho Neto deliberately chose a British national for his protagonist, because during the decade preceding the publication of the novel, England was emerging from the Victorian period to become a site for debates about changing gender roles and sexuality. Among the key events of this era were Oscar Wilde's trials for homosexuality, beginning in 1895, the founding of the National Union of Women's Suffrage in 1897, and the publication of Bram Stoker's *Dracula* (1897), all of which signaled changes in the traditional roles of men and women. In the literary world, the novel *Dracula* is well known for its exploration of latent homosexuality, racial mixture, and women's sexuality.[21] Coelho Neto's novel reveals similar concerns, centered mainly on the mysterious identity of James Marian, whose nature is likened to that of the sphinx because of his queer or dual nature. In the Greek tradition, the sphinx has a lion's lower body, suggesting masculine strength, and a female upper body, often portrayed with breasts, and a female head.

Roberto de Sousa Causo cites Belgian Fernand Khnopff's 1896 painting *The Caresses* as a source of inspiration to Coelho Neto's *Esfinge* (*Ficção* 115–16). In the painting, the sphinx is visualized as a type of femme fatale, a creature that is a portent or warning of things to come, containing both the fearsome and the sublime or otherworldly. At times, Coelho Neto's protagonist evokes Kristeva's "abject," that which provokes a visceral sense of horror and revulsion, like that brought on by bodily excretions (2–3). At the same time, the fascination with the abject often makes it an object of curiosity. According to Judith Butler, this sense of repulsion/ fascination may extend to the social practices of certain groups whose rejection consolidates hegemonic identities (*Gender* 181–82). Although Coelho Neto's James Marian sometimes experiences this type of social treatment, his wealth generally insulates him from outright rejection, while his connections with the occult sciences afford him otherworldly powers.

In *Esfinge*, the unnamed narrator, a writer and translator who resides in the boarding house, discovers James Marian's secret when he is asked by the Englishman to translate his diary. From this embedded text we learn that James was orphaned at an early age and taken to an estate to be raised by a housekeeper and a mysterious man named Arhat. At age thirteen, James gains the company of a boy and girl and begins to fall in love with the girl. However, after Arhat suddenly dies, his departing spirit visits James and reveals a great truth to him, causing him to fall into a three-month coma. It seems that some ten years earlier, Arhat, having learned of a terrible accident involving a brother and sister, had used his surgical skills and the arcane arts (*magna scientia*) to save the body of the boy and the head of the girl by making them into a single being—James Marian. Upon awakening from the coma, James breaks off his relationship with the girl he loves, because he has the head and brain of a woman. He represses his sexual desire, and once he comes of age, decides to travel the world.

Roberto Causo explores parallels between Coelho Neto's James Marian and the monster in Mary Shelley's *Frankenstein* (1818), characterizing both as artificial creations abandoned by their creators, although to be fair Arhat does attempt to offer guidance to his ward (*Ficção* 113–15). I believe that a more fruitful comparison can be drawn between James Marian and Bram Stoker's Dracula. Like Dracula, James attracts men and women alike, although his contact with them is mainly through conversation and unrequited love, while Count Dracula bites the necks of his female victims, suggesting penetration and sexuality. Although not he is not predatory, James Marian ends up seducing men and women alike with his attractive appearance.

In fact, James's seduction of his "victims" is almost inadvertent. The first, the teacher Miss Fanny, falls in love with him after listening to music with him in the garden. When James decides to spend more time away with a male friend, Miss Fanny's tuberculosis dramatically worsens and she dies soon after. While the lower or male part of James's body is attracted to Miss Fanny, he cannot bring himself to take the relationship beyond a flirtation, since his head is female and the relationship would be tantamount to lesbianism.[22] Notably, Miss Fanny's tuberculosis causes her to cough up blood, which connotes sexuality and violation, as if she were condemned to die for loving this queer being.

At this point in the story, the text brings in elements of the supernatural as we learn that James is able to produce ghostlike projections of himself. Near the time of Miss Fanny's impending death, James casts a mysterious spell on her and her fellow boarders, appearing before them magically, dressed in a tunic, while he is actually in another part of Rio interacting with others. Before she dies, Miss Fanny has another vision of James, and the narrator himself states that he saw "James Marian e, naquele traje, o seu rosto realçava mais belo. Era ele como eu o imaginara em devaneio" (85; James Marian, and, in that outfit, his face even appeared more beautiful. It was how I imagined him in my daydreams). Here the narrator admits to seeing James's incredible beauty as if he had imagined him in a daydream, suggesting an almost sexual fantasy. Another boarder, the musician Frederico Brandt, whose piano music charms both James and Miss Fanny, also confesses to having the same vision of a tunic-clad James floating in and out of sight near the boarding house. However, the musician likens James to an angel, a genderless being, emphasizing his androgynous aspect. Miss Fanny's death is real, however, which leads us away from the spirit and back to the body and the problem of gender identity.

Even though James rejects his female admirer, he forges an intimate bond with the male narrator, who, as the translator of James's diary, fully knows his secret identity. Toward the end of the narrative, when James abruptly decides to leave, the narrator recounts a scene in which he dutifully returns the original manuscript to its owner. The scene would be banal were it not for the fact that later we learn that James had already left by ship the day before, making it seem as if the narrator hallucinated or experienced an uncanny, supernatural

event of James's return. After learning about James's earlier depar-
ture, the narrator suffers a breakdown from which he recovers only
months later, when he awakens to find himself in an asylum. When
a former boarder (and medical student) visits the narrator there, he
offers a "scientific" explanation for the narrator's experience: neurasthenia, the all-purpose malady of the over-stimulated urban dweller.
Here the narrative offers us a rational solution, leaving us with a
Todorovian hesitation between the natural and the supernatural. This
hesitation, I would argue, represents the space in-between where the
topic of queer sexuality slips in.

On the one hand, the novella's message seems to be to warn
us away from the queer or the occult. Unlike Rafael Antigas of *El
donador de almas*, the narrator of *Esfinge* has not fully inoculated himself against the effects of the arcane arts, and suffers a breakdown.
We can extrapolate that, while Fanny experienced physical death
because of her desire for James, the narrator undergoes a type of
psychological death or insanity brought about by his desire for the
mysterious, sexually ambiguous protagonist. James brings up issues
of repressed homosexuality and bisexuality as a threat to heteronormative society by a foreigner. It seems that contact with James Marian evokes fears of degeneration that could undermine the values of
Brazilian society at the turn of the century.[23]

James Marian leads a privileged existence and is able to control his
own life circumstances, and while his queer body may be perceived as
a threat to some, his presence effects a spiritual revelation in others.
Coelho Neto may have been influenced by Allan Kardec's spiritism,
which found fertile ground in Brazilian society, where syncretic religious practices, popular Catholicism, and Comtean positivism were
popular among the middle and upper classes. The declaration of the
Republic and the separation of church and state allowed Kardec's
doctrines of healing, charity, reincarnation, and karmic evolution
to flourish (Hess 14–16) and to become a source of heated debate in
Brazilian society as a source of psychiatric therapy or psychic harm
(Moreira-Almeida 9).[24]

In this light, James Marian becomes an even more enigmatic character, since one of the more surprising elements of Kardec's doctrine

is his claim that spirits do not have gender (Hess 77). James Marian's creator, Arhat, is from India, which connects him with the arcane arts and Eastern religions. As De Cicco notes, the occult provided new types of knowledge for artists and seekers of knowledge beyond material science. In Coelho Neto's tale, the narrator/writer survives his ordeal with James Marian and is left queered, having broken down heteronormative binaries and recovered from his encounter with the irrationality of the occult. Although James Marian fails to find a sympathetic community in the Brazilian boarding house, he served as an inspiration to its artists—the musician, the teacher, the writer/translator. If physical contact is taboo for James Marian—as shown by his reaction to Fanny—spiritual evolution is still an option. Thus, James Marian's spiritual or otherworldly impact may reflect what Halberstam has called a "de-linking" from the "organic and immutable forms of family" (The Queer Art 70) represented by artists and others open to new forms of knowledge, beyond the traditional values of the Oedipal home and heteronormative society.

After these turn-of-the-century examples, the theme of women experiencing a man's body all but disappears in both Mexico and Brazil during the period from 1930 to 1970.[25] The next significant work in this tradition is Roberto de Sousa Causo's 2008 novella, O par: Uma novela amazônica (The pair: An Amazonian novella), in which the male protagonist experiences the body and thoughts of his female companion, while she gains his physical capabilities. The experience leaves them both ambiguously gendered, both psychologically and physically.

While Causo's O par involves psychological trauma, it also brings up a different type of reproduction akin to cloning but with a twist, because it involves aliens who first clone and then recombine human DNA and memories regardless of race and gender. To help himself recover from the death of his lover, Joana, the victim of a fatal car accident in São Paulo, Oscar Feitosa enlists in the army and is sent to the Amazon where a mysterious war is being waged. Once there, he learns that aliens have killed most of the local population, leaving their dehydrated bodies behind. Feitosa begins to distrust the motivations of his superiors and comrades, deserts his unit, and kills anyone who threatens him. He meets Conrado, a character reminiscent of Kurtz from Joseph Conrad's *Heart of Darkness* (1899), who has legal claims to the land. In tacit agreement with the military, Conrado is using weaponry and henchmen to enslave workers and poachers in order to establish his own personal empire, from which he plans to increase his sales of animal pelts and drugs to foreign markets.

In his travels through the Amazon, Feitosa learns that the aliens do not always kill their human victims; instead, they use them for bizarre experiments, extracting pieces of leg tissue and memories from their brains in order to recreate or clone their dead loved ones, who become inseparable from the human hosts they originated from. Feitosa meets several of these pairs during his travels, some of whom fall victim to Conrad's henchmen, who suspect that they are were-wolves or humans implanted with tracking devices by the military. After wandering for days, Feitosa is rendered unconscious after an encounter with aliens. He awakens to find Joana alive at his side. However, she is now white like him, instead of black, and is endowed with his physical abilities as a soldier.

These impossible pairs haunt the Amazon landscape as victims of displacement and symbols of environmental destruction. Feitosa realizes that he has also been turned into a pair with Joana, and their fates and consciousness are tied. Although his grief at Joana's death is relieved when she appears by his side, his trauma is renewed when she is later raped by Conrado's henchmen. Because they are a pair, Feitosa shares Joana's thoughts and feelings, a sensibility that becomes more acute after this traumatic event. Finding himself no longer able to operate within the conventional boundaries of masculinity and femininity, he becomes, as Judith Butler has observed, an example of "beings who do not appear properly gendered; it is their very human-ness that comes into question" (*Bodies* xvii). When Feitosa realizes that Joana may only be a projection of himself, he begins question-ing his male sexual identity and masculine power. In this ambiva-lent state, he seems to embody Butler's deconstruction of Lacanian theory that "opens up anatomy—and sexual difference itself—as a site of proliferative significations" (*Bodies* 56), undermining the assump-tions of heteronormative sexuality and power. As he internalizes his new sensibility, affecting both mind and body, Feitosa realizes that his "masculinist and heterosexist privilege" has been queered (Butler, *Bodies* 56). He can no longer operate within the parameters of masculinity, because after the rape, he experiences what Butler has called the "lesbian phallus," or a signifier that is "significantly split, for it both recalls and displaces the masculinism by which it is impelled" (*Bodies* 56). This disassociation of power and masculinity is profoundly disturbing and Feitosa can hardly face Joana after this event, perhaps more out of his own loss of agency than any sense of shame.

Metaphorically, these pairs are eerie reminders or living representations of traumatic memories, outside of time and the "locus of referentiality" (Caruth 6). As Cathy Caruth explains in her 1996 study *Unclaimed Experience: Trauma, Narrative, and History,* victims of trauma have a simultaneous desire to remember and to forget, so the trauma narrative is an attempt to capture both the life of the survivor and the memory of the loved one, implying both the impossibility and the necessity of a "double telling" (8), here outwardly represented by the pairs. As strange, unacceptable products of alien science, the pairs find themselves adrift in society. Joana and Feitosa find themselves heading north, toward what, they imagine, is the location of the alien ship. Rejected by human society and its categories, the pairs cannot fight for their rights or directly confront either the aliens or the humans. Only by accepting their queer experience and their trauma can the pair become part of the baroque ethos, seeking survival either in a new society deep in the Amazon or aboard alien ships, that is, in a society that would reconfigure itself to accept nonheteronormative pairings.

As we have seen in *Donador de almas, Esfinge,* and *O par,* gender fluidity, failure, escape, and evasion are indicators of the baroque queer. Queered spaces become the spiritual or alien realms where the female side of the pairings (Alda, Marian, Joana) escape to resist patriarchal societies. We find that professions (governance, medicine, the army) and places (the boarding house) generally reserved for men are queered in subtle ways. Yet each experiment to share this space fails in society, so participants find a third way, one that produces a distinct version of *codigofagia* of gender, mixing and remixing as a way to resist dominant paradigms of conventional pairings and foundational fictions.

Women as Other Species

James Tiptree Jr. (pseud. Alice Sheldon) was a feminist author of science fiction who used a male point of view to illustrate female invisibility in the 1973 story "The Women Men Don't See," a story that typifies how women survive on the margins of patriarchal society in a manner similar to the baroque ethos.[26] After a plane crash in Quintana Roo, a mother and daughter pair opt to try their luck with aliens rather than remain with the arrogant male narrator, their self-professed rescuer. Tiptree's story is just one example of female alienation in science fiction and fantasy: women also opt out of their predicament by identifying with other species of animals, including amphibians, reptiles, and marine life. Interspecies identification functions as resistance to patriarchal or capitalist definitions of femaleness and queerness.

Mexican author Efrén Rebolledo's "Salamandra" (1919; Salamander), in which a woman becomes identified with a salamander, is a fascinating work of proto-feminism. Although it is less well known than Eduardo Urzaiz's masterpiece of early Mexican science fiction *Eugenia: Esbozo novelesco de costumbres futuras* (1919; Eugenia: A fictional sketch of future customs), "Salamandra" reflects similar anxieties of gender and reproduction that arose in Mexico after the 1910 Revolution.[27] However, "Salamandra" is also of interest due both to the connection it makes between women and other species and its links to the queer and the horror genre through its reference to Stevenson's *The Strange Case of Dr. Jekyll and Mr. Hyde* ("Salamandra" 272).

Although not explicit, queer desire comes to the surface in Rebolledo's story when the protagonist Elena Rivas begins to spend time with a female companion, "con quien salía a todas partes" (265; with whom she went everywhere), including parties, literary events, and a weekend at her family's country home. Although she is desired by men and is courted by León, Elena rarely consummates her relationships with them, but rather uses them like mirrors to reflect her own power, getting more pleasure out of voyeurism than actual physical contact (Barrera Barrios 98–99). She is compared by the narrator to several abject creatures, including salamanders and spiders, that suggest unnatural parasitical or nonhuman reproduction associated with cold-blooded or blood-sucking creatures. The metaphor of Elena as salamander is key to the interpretation of the story, especially in the way she uses and discards her lover, León.

Elena's unnatural beauty, financial independence, and intellectual interests make her a powerful and threatening presence. Although born in Mexico, Elena was educated in Los Angeles where she "adquirió esa independencia que distingue a las mujeres de los Estados Unidos" (261; she acquired that independence that distinguishes women from the United States), which seems to be part of her monstrosity. She is described as a female dandy, a traditionally male role she regularly adopts at intellectual soirées and debates, and male intellectuals are among her prey. Her androgynous salamander body becomes a kind of shape-shifting vehicle used to question gender roles by transforming the social environment. The creatures associated with Elena engage in a mode of reproduction that estranges us from human and traditional female roles. Salamander and spider females lay eggs generally outside the body, unlike mammalian species. The images of egg chambers in films like Ridley Scott's *Alien* (1979) and its prequel *Prometheus* (2012) are reminders of the abjectness and horror of external fertilization and asexual reproduction. Elena's trail of former lovers, the incestuous passion she provoked in her brother, and the death of her first love and subsequently of her first husband are typical of the archetypal femme fatale. Her androgynous body is seen not as fertile but fatal, very different from the reproductive body prescribed by the eugenic precepts of the period, which took on special urgency after the devastation of the Revolution and subsequent political unrest. Léon, her putative lover, becomes the victim of her wiles.

The connections to the Revolution become clear when we consider the streets and locations in the text, because by referring to the country's historical narrative, they reinforce the body/society allegory. In Mexico City, Elena lives in a hotel in the neighborhood of

Colonia Juárez, a reference to the revered nineteenth-century president who died during his second term in office in 1872. The hotel is located on the Avenida Francisco Madero I, named after a leader of the 1910 Revolution who later became president and was assassinated in office. Elena's lover, León, also walks along the Avenida Independencia, whose name memorializes Mexico's struggles for independence from Spain from 1810 to 1822. This street, in turn, leads to Elena's countryside home near Querétaro, where Maximiliano, the ruler of Mexico during the French occupation, was executed in 1867 by order of Juárez. In Rebolledo's story, Querétaro, with its depressing, dry, and desolate atmosphere, is a place where calves are born dead and the land is sterile, factors that drive Elena and her female companion back to the city after a weekend there (274).

Elena's androgyny and her history of betrayal can be read as harmful to Mexican nationalism, since she, like the landscape of Querétaro, is barren. Her betrayal of the ideals of the 1910 vision of a fertile and productive Mexico is effectively captured by the salamander metaphor, since these animals poison their prey. They have also long been associated with newts, witches, and the occult.[28] Uncontrolled by the patriarchy, Elena exemplifies the phallic female, a figure that, as Barbara Creed notes, helps us understand the social and psychological conflicts evoked by the figure of the "witch, archaic mother, monstrous womb, vampire, *femme castratrice*, castrating mother. They shock and repel, but they also enlighten. They provide us with a means of understanding the dark side of the patriarchal unconscious" (165–66). As a woman who assumes predominantly male values, Elena illustrates the fears that surface when the patriarchal control of motherhood and reproduction is threatened.

Amphibians and reptiles take on a different connotation for Diana Tarazona in her novel *El animal sobre la piedra* (2008; The animal on the rock), which deals with oviparity, a form of reproduction that allows for maternity outside of traditional family structures, a new type of female body, not only queered but trans-species.[29] The story's protagonist and narrator, Irma, presents herself as self-engendered, separate from traditional families and reproduction. While Tarazona's novel belongs more to the genre of the fantastic than to science fiction, it could be compared to Kafka's *Metamorphosis* (1915) given the protagonist's bodily transformation. However, the transformation depicted by Tarazona is different from Gregor Samsa's, because readers may question her corporeal change, bringing up Todorov's theory of hesitation between the natural and supernatural in fantastic texts. While reading the novel, we hesitate because of the apparent naturalness with which Irma accepts her reptilian metamorphosis.

As in many fantastic texts, Irma is not completely reliable as a narrator, since she is numbed by the grief she feels over her mother's death. She flees to a seaside resort, where she undergoes a transformation: her skin becomes scaly and her hands and feet become like those of a lizard. In the neutral tone used throughout the narrative, she comments on how she has started to spit poison and to inseminate herself like a frog. After a while, she prepares to lay her egg near the bank of a stream, where she feels safe. Bit by bit, images of a hospital—or possibly a mental hospital— begin to filter into the narrative.

At this point a detail that Irma mentioned at the start of the narrative falls into place. She has a mark on her wrist from an intravenous drip, which allows us to reformulate the narrative. A nurse mentions that Irma was found wandering naked, a fact Irma neither confirms nor denies. She appears to be relating normally to the doctor and nurse about her physical condition, but suddenly interrupts the conversation saying that she has left her egg on the chair, asking that it be placed in the sun for warmth. At the end of the novel, Irma appears to crouch and shrink, seeing the world from the perspective of a lizard. Looking under the bed, she finds the egg. However, when she touches it, she finds that it is hollow.

As to the significance of this hollownesss, we can imagine several possibilities. It could be the result of a depression provoked by her mother's death, a sense of guilt, or a miscarriage, or it could be a psychological emptiness that she cannot manage to fill. At one point, Irma makes a reference to the oviparous nature of frogs, thinking about how, as a child, she would hunt tadpoles with a friend. She remarks, "Estábamos seguras, sin que supiéramos por qué, de que esos renacuajos vivían solos, nunca pensamos en sus madres. Pero las ranas no regresan a ver a sus hijos: teníamos razón" (152; We were sure, without knowing why, that these tadpoles lived on their own, we never thought about their mothers. But frogs don't return to see their offspring; we were right). The psychological independence of the tadpoles is what she longs for, because she is still trying to free herself of the feelings provoked by the death of her mother. What Irma really desires is to be freed of the pain of guilt and loss; she longs for the cold bloodedness of amphibians and reptiles. She asks herself, "Si estoy transformándome en un reptil, ¿mi descendencia será ovípara?" (80; If I am transforming myself into a reptile, will my offspring be oviparous?), as if she wished to assume this new way of being or of understanding her body. When the nurse refers to her placenta (168), we realize that Irma has perhaps undergone a miscarriage or stillbirth, making us even more doubtful about what has truly happened to her.

The female oviparous reptile body might also represent a physical, emotional, or intellectual process of change. The desire to redefine oneself after a trauma is captured in the contrast between the scaly body of the reptile and the purity of the white egg. Outside of the body and the family, the egg allows the woman an almost male sense of gender performance, of independence and autonomy, while also participating in birth and creativity on her own terms, without excess sentimentality or melodrama, imagining alternatives for the female body and physical power. However, the hollowness of the egg returns us to a sense of loss or existential crisis. This failure typifies the baroque ethos and a refusal to conform to traditional images of motherhood, such that the trauma of remembering and forgetting are captured in a queer narrative of failure, where fissures and questions remain.

Aline Valek's 2016 *As águas-vivas não sabem de si* (Jellyfish know not of themselves), another trans-species novel, is about a Brazilian diver, Corina, who works with a small submarine crew testing a new deep-sea pressure suit to be used by a large corporation. The text has many lyrical or baroque descriptions of aquatic life told from Corina's point of view and that of other characters. As miscalculations and accidents plague the voyage, Corina begins to identify less with the crewmembers and more with marine life, whose interdependence she admires. This sensibility evolves throughout the narrative in descriptions of the subjectivity of underwater creatures, including a curious octopus, a sperm whale, a humpback whale, a jellyfish community, and the azúlis—an extinct underwater civilization. All of these reinforce a sense of empathy and identification with a great diversity of beings.

Although, as noted by Rüsche and Furlanetto, the sub-themes of the novel include ecological devastation as well as struggles of class and gender (263), I find Corina's personal and cross-species relationships to be of equal interest. All five crew members have secrets that rise to the surface once the stress of the underwater mission takes its toll on them, and they begin to grow increasingly isolated and suspicious of each other. Hierarchies of gender, nationality, class, and education create conflict, as the novel begins, between the black American male professor, Martin, and the Brazilian female engineer, Susana. The Brazilian divers, Corina and Arraia, are lowest in the hierarchy, treated as outsourced labor, while the professor's Brazilian doctoral student, Maurício, fits somewhere in between. These social differences reinforce the novel's overarching theme of existential solitude, a sentiment repeatedly expressed by Corina, as noted by Matangrano and Tavares (178). I would add, however, that the human desire for a sense of belonging that recurs throughout the novel is reminiscent of Santiago's sense of queerness and community and Gazi Islam's anthropophagic sense of the other.

The sense of the other is cultivated by the idea of communication and listening that recurs during the narrative. The ship's name, *Auris*, means "ear" in Latin, and the role of aural data is key in the novel, since the crew must listen and work together after a mysterious technical breakdown leaves the crew without light and oxygen for six minutes until Susana, the engineer, manages to get the system online. The outage erases Martin's invaluable data showing that his whale recordings evoke responses from intelligent underwater beings he has attempted to contact. Corina manages to connect in a meaningful way with all the crewmembers, telling Susana that her capabilities as a captain saved them, expressing confidence in Maurício's data, and helping Arraia recover from a psychological breakdown. She supports Martin's research and manages to turn the team into a functioning unit after a devastating fight. Although she is at the bottom of the social hierarchy, Corina convinces them to proceed with the data-gathering mission, despite its inherent risks.

In a recording made before leaving on her last dive, Corina observes that all of the crew members are "broken" or have "missing pieces." By the end of the novel we learn that Martin fears losing his reputation due to his theories about the existence of an underwater alien civilization, while his assistant Maurício fears being outed as a homosexual. Susana suffers pangs of guilt for the death of a colleague that took place under her command, and Arraia is hiding the fact that he is a recovering alcoholic. He is the only one to whom Corina reveals her own secret, that she has been diagnosed with multiple sclerosis.

Despite her illness, Corina listens to her body, enjoys the feeling of the vibrations of the whale's song, relies on her skills as a diver, and always allows her body to dictate her actions. While on the last dive with Martin to plant the recorder that will capture crucial data, she encounters an intelligent life form known as a specter (*espectro*) and feels the urge to follow it, such that she detaches her breathing tube as Martin attempts to pull her back. We are left to ponder her action as an expression of either despair or liberation.

Corina's sense of existential solitude haunts her throughout the novel, and it is echoed in the fate of the jellyfish that give the book its title. In one chapter, the jellyfish is described as a glorious whole whose collective expanse captures light, dazzling those aboard a cruise ship. Yet when the ship cuts through the colony, its members are set adrift, and one ends up on a beach, sensing separation, despair, and its imminent demise. The subjectivity attributed to the ocean creatures and their ecosystem suggests a sense of collectivity longed for by characters who strive for a sense of an oceanic consciousness. While identity politics separate the characters at times, Corina manages to make the crew whole, absorbing, including, and developing connections, a manifestation of the anthropophagic queer. The community she forms reaches beyond heteronormative models, evincing the baroque ethos's strategy to resist traditional definitions of power and individual success, dissolving into an oceanic awareness. Her death dive into the depths of the ocean is another example of the queer art of failure and is equally poetic in its longing.

Conclusion

The women warriors portrayed in this chapter use baroque strategies that appear unproductive or nonlinear: passivity, evasion, and sleep. In Machado de Assis's Orientalist "As Academias de Sião," the concubine Kinnara's subjugation is similar to that of slaves and women on whose productive and procreative bodies Brazil's largely agricultural and patriarchal society depended. Machado's work is suffused with irony, questioning the basis of power, and Kinnara's situation as a philosophical, pregnant former woman warrior is one of its emblems. The Amazonian society described in *A Amazônia misteriosa* queers gender roles, creating a noncapitalist, nonheteronormative society. The protagonist's experience of this new world causes estrangement or queering that compels him to question gender, science, and the colonial practices that are the pillars of his society. In *Duerme*, Carmen Boullosa's sixteenth-century woman warrior inscribes race and gender into Mexico's history of the Conquest. She displays a type of "corporeal fluidity" (Cohen 19) to resist and question narratives of heroic battles. While foundational fictions allegorize violence toward women that formed the basis of mestizo nations, queering them as women warriors uncovers the baroque ethos of resistance.

Women in men's bodies queer homosocial spaces reserved for men, engaging in strategies of evasion to avoid conflict. When Amado Nervo and Coelho Neto introduce female subjectivity into male bodies, they infiltrate worlds associated with male privilege. The men are queered yet survive the experience, while the women disappear or perish. This violence comes to the surface in Causo's *O par*, in which cloning—carried out by alien engineering—breaks with the idea of harmonious *mestiçagem* as an ideology. Joana's rape is a reminder of the history of sexual violence hidden within the ideology of *mestiçagem*. It is significant that this comes light in the Amazon, the repository of the nation's collective unconscious.

Women identified with other species question heteronormative values of reproduction, nation building, and modernity. Rebolledo's salamander-like Elena plays the role of the devouring, sexually ambiguous creature associated with the violence and crisis of national values during the period of the Mexican Revolution, a time that saw independent women as threats because of their refusal to conform to their reproductive roles in *mestizaje* and the rebuilding of the state. In Tarazona's narrative, an egg (her offspring) is Irma's last hope, yet when she picks it up, it is hollow. While this may represent her pain and resistance to clichés of miscarriage, mourning, and loss, when read as a national allegory, it symbolizes the failure of the Mexican state and its supporting ideology of *mestizaje*. In Brazil, Aline Valek's character identifies with marine life that embodies her sense of solitude and a desire for community. After reconnecting the crew, her decision to return to the deep sea (and commit suicide) is a personal decision as well as a statement about the planetary destruction we are currently witnessing.

In these Mexican and Brazilian examples of speculative fiction spanning over a century, we see how sociopolitical aspects of the queer, which are often represented as corporeal transformation, resist canonical formations of *mestizaje* and the heterosexual basis of foundational fictions through the baroque ethos of resistance and the queer art of failure, which allows protagonists to ignore conventional definitions of success and liberation, engaging in "refusal, passivity, unbecoming and unbeing" (Halberstam, *The Queer Art* 129). Their stories reimagine social relations and society, reformulating and anthropophagically absorbing the other, be it male, female, nonbinary, gay, alien, cloned, amphibian, reptile, or marine mammal. Halberstam posits that "Queer studies offer us one method for imagining, not some fantasy of an elsewhere, but existing alternatives to hegemonic systems" (*The Queer Art* 89) in utopian spaces where we can rethink definitions of advancement and resist capitalism's definitions of success. For Gazi Islam, "The corporalized nature of anthropophagy, finally, contrasts with the identity politics of some postmodern discourse" (173). The baroque ethos and its advocacy of *codigofagia* seek to recombine and reimagine bodies as resilient, eccentric, and imaginative forms of gender that craft new tales of nation and community.

CHAPTER 3

Trauma Zombies, Consumer Zombies, and Political Zombies

No study on the body in speculative fiction would be complete without chapters on zombies and vampires, which are among the monsters that most frequently haunt our present. As creatures that combine the living and the dead, on the borderline between binaries, they often represent a sense of collective and personal trauma. While each conveys a distinct fear of the contagion, destruction, and the breakdown of categories, they are also portents, as Franco Moretti has noted: "The monster expresses the anxiety that the future will be monstrous" (68). I turn my attention in this chapter to the zombie as portent and embodiment of fear and trauma.

As an element of the increasingly apocalyptic scenarios of the twenty-first century, the zombie figure has been used to presage economic, political, or social chaos in a variety of contexts over the past hundred years, including Mexico and Brazil. The Latin American zombie illustrates Echeverría's baroque ethos both as a manifestation of and resistance to capitalism, an economic system whose spread can be conceptualized as viral. I will argue that this global economic system affects all classes, such that both the dominant and the subaltern are infected and transformed by its relentless spread.

In his 2018 article on cannibal resistance in Latin America and the Caribbean, David Dalton conceives of zombies as an extension of the tradition of cannibals, creatures that have a long history of resisting oppressive figures through the simple expedient of devouring them ("Antropofagia" 14), and I would argue that Echeverría's baroque ethos—the historical result of racial and cultural strategies of survival—provides a basis for this "resistant zombie." Dalton notes that the zombie/cannibal "often metes out justice to those who have enabled the suffering of the colonized masses of Latin America and the Caribbean" ("Antropofagia" 6). He traces the movement of the cannibal figure from the periphery, as the barbarian outcast in the *barbarie/civilización* debates of Latin American identity, to the center stage as the resistant subaltern in the twentieth and twenty-first centuries, that is, from Brazilian Oswald de Andrade's cultural cannibal in his 1928 "Manifesto Antropófago" to Cuban Roberto Fernández's political cannibal in *Calibán* (1973) and finally to contemporary zombies in Latino and Latin American literature ("Antropofagia" 2–5).

As mentioned above, literary zombies in Mexico and Brazil are sometimes used to portray the struggle within the social body against the infection emanating from global capitalism and the internal political dissent it engenders. As Bolívar Echeverría posits, Latin America has been a part of capitalist modernity since its inception, yet as a peripheral region dependent on the sale of commodities, it has been subject to a large degree to the vicissitudes of external markets in the export economy. If the baroque ethos is a system for survival of and resistance to capitalism, then it seems natural that Latin American zombies would fit into this immunological paradigm. Economies that are dependent on commodities are vulnerable to the corrosive aspects of capitalism the way the body is to viruses. In the first part of the chapter, I examine humans who become zombies because of unexpected economic shifts that destroy a way of life. I call these iterations proto-zombies because they are living humans who act like zombies due to traumatic and disruptive events. Although it might be assumed that Latin American literary cannibals and zombies would be associated solely with subaltern groups, I will show that zombie figures represent a variety of social classes. Another category of proto-zombie narratives that begin to appear in the 1940s features characters who are affected by mindless consumerism. Such tales tend to be more satirical in nature, functioning as critiques of economic paradigms of capitalist productivity.

In these first two categories, the Latin American cannibal/zombie serves as a metaphor of infection from external economic systems. Paradoxically, I have found that these figures can also project a "counter-immunological" resistance when populations find themselves targeted by state or corporate control. I will therefore have occasion to use the concepts of both zombie-as-contagion and zombie-as-resistance to interpret Mexican and Brazilian stories that can be seen as original case studies in class struggle and historical paradigm shifts throughout the twentieth and twenty-first centuries.

An Overview of Zombie Criticism and Typology

Since the zombie is a widely known figure and has diverse iterations, it is important to review the history of this figure, which I would define as an active, reanimated human in the broadest sense, that is, an exemplar of the "living dead."[1] According to most scholars, the zombie made its way into the American imagination after the US occupation of Haiti from 1915 to 1934 via William Seabrook's *The Magic Island* (1929), a travelogue in which he first described Haitian workers as zombies in his chapter ". . . Dead Men Working the Cane Fields" (93). Subsequently, the image was popularized or transformed by Hollywood movies such as *White Zombie* (1932), *Outango* (1936), and *I Walked with a Zombie* (1943), which can be interpreted as reflecting American anxieties about race and sexuality during this period. As Ann Kordas states in her study, racial fears associated with the health of the social body played out in the treatment of female sexuality in the 1930s and '40s, because zombie transformations conveyed the idea of punishment for a possible racial or immigrant invasion of the body politic. In the 1950s, the "pod people" of Don Siegel's film *Invasion of the Body Snatchers* (1956) can be understood as Cold War zombies, interpreted alternately as conformist suburbanites or communist sympathizers, since in both cases, they lack any spark or sense of individuality (Moreland 81). This film belongs to the subgenre of science fiction horror that also features the giant ants of *Them* (1954) and avian invaders in Hitchcock's *The Birds* (1963), where the idea of swarms or flocks is used to convey changes in social and moral paradigms (Mahoney 122). The fear of mass conformity emerging in the US required new monsters to convey the idea of impersonal corporate or government bureaucracies, and somnambulist-like zombies filled the bill.

Also notable is the issue of how fast zombies move. Zombies amble slowly in the movies of the 1930s, but tend to move more quickly over time. This is apparent in films ranging from the lethargic zombies in George Romero's *Night of the Living Dead* (1968), to the moderately rapid zombies in the same director's *Day of the Dead* (1984), to the alarmingly fast zombies in more-contemporary films such as Zack Snyder's *Dawn of the Dead* (2004), Danny Boyle's *28 Days Later* (2003), and Max Brook's *World War Z* (2006). The rapid, swarm-like movement of the zombies in this last film foregrounds contemporary fears of global viral pandemics, whether biological or cybernetic (Riley 196–97).

The metaphor of the zombie contagion / outbreak can be used to explain anything that spreads without warning, including populism, communism, McCarthyism, consumerism, nontraditional sexuality, or AIDS (Beisecker 198). This illustrates how the zombie can represent any group or movement that has been made a target of the irrational fear of the Other. Zombies can represent an encounter with disaster or with what René Girard, in his *Violence and the Sacred* (1977), characterizes as a crisis of "indifferentiation," where hierarchies and boundaries break down and people are faced with social and physical chaos. Stephanie Boluk and Wylie Lenz make this connection in their study of Girard's essay "The Plague in Literature and Myth," which is also useful in analyzing zombie outbreaks (6–7).

In their 2008 "Zombie Manifesto: The Nonhuman Condition in the Age of Advanced Capitalism," Sarah Juliet Lauro and Karen Embry identify the zombie as the posthuman of consumer capitalism. They outline three basic types of zombies in the Anglo-American tradition: the original zombie tied to Haiti and slave labor, which they denote with the French spelling, *zombi*; the mainly somnambulist zombies of 1930s and '40s Hollywood, for which they use the *zombie* spelling; and the zombies associated with global capitalism and consumerism, for which they use the spelling *zombii*. Lauro and Embry postulate that we now inhabit this third phase, that is, we are mindless *zombiis* of global capitalism, feeding off the products of the global market (93), unaware of our own powerlessness against collective power (98). Lauro and Embry even go so far as to claim that the *zombii* is an unconscious swarm organism that embodies a new type of posthuman (88).

Some critics consider Lauro and Embry's concept of *zombii* to be too vague, "a nearly-empty signifier carrying a few key-but-infinitely interpretable features (strength through swarming, lack of individuality or identity)" (Moreman and Rushton 4). For others, such as Boluk and Lenz in their *Generation Zombie* (2011), this trait is a strength of the zombie metaphor: "what seems to be an essential characteristic of the zombie is its capacity for mutation and adaptation. Just as the zombie resists legal containment, it resists generic and taxonomic containment; it is remarkably capable of adapting to a changing cultural and medial imaginary" (9).

In my analysis of the zombie figure in Mexican and Brazilian speculative fiction, I identify three main types. *Trauma zombies* arise in the wake of extremes, either economic collapse or the rise of criminal networks. *Consumer zombies* reflect the unthinking acceptance of consumer capitalism, especially during the middle of the twentieth century during the period of import substitution, while *political zombies* arise mainly from the necropolitics of authoritarian regimes.

Trauma Zombies

As I have noted above, the zombie metaphor applies to all social classes in Brazil, in part because of economic vulnerabilities. The proto-zombies of early twentieth-century Brazilian literature mark the end of the export economy model that was based on the idea that specialization offered trading advantages, a doctrine widely accepted in the 1880s in Latin America.[2] Despite this, both Mexico and Brazil began to industrialize during this period, with different degrees of success. With the declaration of the Republic in 1889, Brazil's first military government, with support from emerging middle sectors, immediately began to promote industrialization. However, inconsistent tariff polices designed to protect local manufacturing led to economic instability, while the issuance of bonds and paper money caused excessive speculation and the great financial crisis known as the *Encilhamento* of 1891. This crisis contributed to general social turmoil during the rest of the decade and motivated authorities to return to their reliance on coffee exports (Burns 293–95), making Brazil's monocultural export economy again subject to the vagaries of international markets.

Mexico was able to employ tariffs as part of a more consistent economic strategy that resulted in higher levels of growth: between 1880 and 1910, per capita imports doubled in value and exports grew at a very respectable rate, smoothing the transition to a more developed economy and integration with the international market (Ficker 192). Care was taken to integrate the purchase of low-priced imported manufactured goods with those produced locally (206). Unfortunately, at the same time, mining interests and capital were concentrated in the hands of the few, creating an extremely unequal distribution of wealth, drawing protests from rural and urban workers alike. Without a strong entrepreneurial or middle class to act as a buffer zone or propose reforms, these political pressures would grow until they caused the Revolution of 1910 (Skidmore and Smith 227–28).

In Brazil, by 1900 national policies returned to promoting commodities—mainly coffee—whose profits provided wealth to underwrite the industrialization of São Paulo, where available land, immigrant labor, and hydroelectric power fomented the growth of factories (Piletti 138). Coffee overproduction required constant government intervention to stabilize prices, a practice that would last until the Great Depression, when coffee was actually burned to eliminate oversupply and buttress prices (Burns 309–12). Ericka Beckman characterizes the Latin American "novelas de la tierra" as "commodity novels," including Columbian José Eustacio Rivera's

rubber boom novel *La vorágine* (1925; The vortex), in which dreams
of wealth are described as a fever, driving men crazy by the millions
(245). For Beckman, it is "the smell of rubber—a commodity produced for the global market—that conjures the 'insanity of the millions'" (167), illustrating a metaphorical contagion that spreads when
local communities come in contact with outside economic forces.

In Brazil, the connection between coffee and gold takes a fantastic turn in Lima Barreto's 1911 short story "A nova Califórnia" (The
new California), which illustrates the idea of zombie contagion as
the result of shifting economic models. When a foreign scientist/
alchemist named Flamel arrives in a small coffee-producing town in
the state of Rio de Janeiro, his mysterious experiments spark curiosity, especially after he recruits the most respected men in town—
the pharmacist, the local politician, and a military man—to help
him carry out his research. At the same time, local inhabitants are
outraged when they hear that grave robbers have struck at the local
cemetery. The city posts a guard to watch over the graveyard, showing increasing concern about Flamel's experiments.

Later, when Flamel's secret is revealed—that he is able to convert
human bones into gold—the idea spreads like a virus through the
quiet town, suddenly turning its good citizens into unthinking monsters who first turn to the local cemetery in search of bones, then
murder each other to obtain this source of potential wealth. Barreto's
metaphorical zombies do not even show any sentimentality toward
loved ones: in fact, one child convinces his father to hunt down and
kill his mother for the gold she represents. The infection is so complete as to turn the town of three to four thousand into an apocalyptic wasteland in one night, without a single respectable soul left
alive, while the foreign instigator and his cohort escape unscathed.

Beckman has shown that commodity fetishism begins with private
fantasy (215n9), but in the case of Barreto's "A nova Califórnia," the
desire for wealth becomes a collective illness that spreads through
society as a whole. The townspeople's simple daydreams of moving
to Rio de Janeiro, attending a private school, or buying more cattle
turn into nightmares once the virus of wealth infects their minds,
essentially turning them into zombies. For example, a young woman

in the story is at first upset and outraged when she hears about the desecration in the cemetery. However, because she wants to show off her beauty on the streets of Rio de Janeiro to win appreciation and perhaps make a successful match, she concludes that desecration is perhaps not so bad, since the past has little to do with her fate: "Que tinha ela com o túmulo de antigos escravos e humildes roceiros? Em que pode interessar aos seus lindos olhos pardos o destino de tão humildes ossos? Porventura o furto deles perturbaria o seu sonho de fazer radiar a beleza de sua boca, de seus olhos e do seu busto nas calçadas do Rio?" (31; What did she have to do with the graves of former slaves and small farmers? How could the fate of such humble bones spark the interest of her lovely brown eyes? By any chance, should the crime of their disappearance disturb her dream of showing off the beauty of her lips, her eyes, and her bust on the streets of Rio?). This is one of the few explicit connections made with Brazil's history of slaves and rural laborers, whose bones are of no interest to a young woman with hopes set on urban life and better economic status.

Lima Barreto's story presents a microcosm of Brazilian society in the First Republic, using zombie imagery to recall the use of slave labor while anticipating the dangers of consumerism and competition implied by Brazil's transition from a rural to an urban society. Roberto de Sousa Causo included "A nova Califórnia" as the opening tale of his second volume of *Os melhores contos brasileiros de ficção científica: Fronteiras* (2009; The best Brazilian science fiction stories: Frontiers), explaining that its alchemy theme is a precursor to modern science and thus to science fiction. While he interprets the tale as a parable of greed that reveals the fragility of moral values, I interpret it as a demonstration of rural Brazil's lack of immunity to capitalist logic. Indeed, the only survivor is the town drunk who refuses to participate in the hunt for bones since, as an alcoholic, he lacks any capitalistic ambitions and is, ironically, the only one with immunity against the disease. Just as an export-based economy has few mechanisms to keep extreme market fluctuations from affecting its society,[3] Barreto's town has no immunity as the capitalist infection takes hold. As zombies, the townspeople effectively embody and parody the precepts of capitalism, carrying out their "work" with an ironic ethos of efficiency that results in the utter destruction of the community.

Through the viral metaphor, Barreto not only illustrates the devastation of newly introduced neocolonial economic structures but also inverts the long-held notion that Europeans are unable to live and thrive in the tropical climate. As Dain Borges notes in his 1993 article "'Puffy, Ugly, Slothful and Inert': Degeneration in Brazilian Social Thought, 1880–1940," pseudo-scientific notions of racial degeneration posited that Europeans became morally degenerate when exposed to tropical climates. This is rendered in literary form in Aluisio Azevedo's 1890 novel *O cortiço* (The tenement), in which the hardworking Portuguese immigrant is seduced by a woman whose tropical sensuality is presented as the cause of his subsequent lassitude and moral decline (Borges 244–45). As an Afro-descendant, Lima Barreto reverses the paradigm; in "A nova Califórnia," it is a European who fatally infects the Brazilian residents with an economic disease against which they have no immunity.[4]

Turning to the state of São Paulo, where much of Brazil's coffee was produced during this period, we find another interpretation of the zombie figure resulting from shifting economic paradigms. I mention this short story because it is emblematic of commodity fetishism and illustrates that proto-zombies can be from the Brazilian upper classes, who ironically become victims of the system from which they once profited. As the son of a plantation-owning family in São Paulo and hence a member of the social elite, author José Bento Monteiro Lobato takes a perspective different from that of Lima Barreto, an Afro-descendant from Rio de Janeiro's popular classes. Lobato's early stories, published in the 1919 collection *Cidades mortas* (Dead cities), refer to the slow death of small towns in the coffee-growing region of the state of São Paulo (Vale do Paraíba) during the first decades of the early twentieth century.

In Lobato's story "Café, café" (Coffee, coffee), which first appeared in 1900 (*Cidades mortas* 182), we witness the zombification of a coffee plantation owner, Major Mimbuia. As a landowner, he is not the drone worker commonly portrayed in zombie literature but a rich man whose one-track mind makes him unable to adjust to the new

economic reality of falling coffee prices. Ignoring all the rational arguments and alternative crops suggested by neighbors and other farmers around him, he persists in believing that the price of coffee will
rise. After losing money, workers, and his crop, he is finally forced to
sell off almost all of his land and the family home. In the end, he is
described as a zombie-like being, "já nu de todo, os olhos esbugalha-
dos a se revirarem nas órbitas com desvario" (182; all but naked, with
bulging eyes rolling in their sockets), whose frenzied work is that of
"um possesso" (181; one possessed). The coffee plants on the small par-
cel of land that he retains are infested by invasive weeds that seem
to feed off of the major, turning him into a zombie. As an obsessed
man, with his brain almost gone, he joins the ranks of the living dead,
a trauma zombie ironically similar to the figure of the zombie-slave.

The power of Lobato's story lies in its compressed sense of time:
years pass in a matter of paragraphs, as we watch the character transform from an arrogant planter into a metaphorical zombie. His transformation captures how an entire social class with a long, prosperous
history was destroyed in short span of years, as profitable agricultural
communities turned into ghost towns or *cidades mortas*.

Asynchronies, uncanny pauses, and what Terry Harpold has called
"zombie time" play a role in both stories.[5] The zombie trope implies
an interruption of the expected course of accumulation and progress,
marked by a sudden economic or technological change that suspends
the sense of continuity. Harpold feels that zombies reflect a horror of
emptiness and anticipation of shifting paradigms or the feeling that
one is experiencing a sense of being "neither living nor dead, but somehow in between both orders" (161), similar to the fate of the former
coffee planter. Thus, the fast viral contagion in Lima Barreto's story
about the ghoulish invasion of the cemetery contrasts with Lobato's
slower- paced story about the addle-brained coffee baron. Yet when
considered together, they illustrate the discontinuous alternation of
fast and slow, capturing the unsettling transition between rural and
urban rhythms of life as Brazil reacts to changes in commodity markets
and shifting economic paradigms, against which it has no immunity.[6]

Economist Douglas H. Graham notes that, during this period,
despite unjust patterns of labor exploitation, Brazil's culture of

accommodation led to a stabilization of patterns of inequality in such a way that massive outbreaks of social unrest failed to occur, despite government violence against regional revolts (14–15). Brazil's degree of social unrest cannot compare with the scale of the Mexican Revolution, which had over a million casualties.[7] In other words, while the stories by Lima Barreto and Lobato portray the anguish of economic displacement in Brazil, the outbreak of violence in Mexico was literal and not metaphorical. Despite its success in modernizing its economy, Mexico's growth was far more "exclusionary, exploitative and foreign dominated than the Brazilian coffee boom" (Graham 15–16), and what began as a constitutional challenge by Francisco Madero against the protracted rule of Porfirio Díaz soon became an out-and-out class war. The Mexican Revolution is a complex struggle of alliances and interests that would last nearly ten years as politicians such as Madero and Obregón joined forces with popular leaders, including Pancho Villa, Emilio Zapata, and Pascual Orozco, against government troops known as *federales*.

The living dead are not entirely absent from Mexican literature during the early part of the twentieth century. I agree with Carmen Serrano's idea that that the figure of the proto-zombie is suggested by José Guadalupe Posada's engravings or drawings of *calaveras* or animated skeletons ("Mapping" 462). These cartoonish figures appeared in the "penny press" or posters known as broadsides that were marketed to Mexico City's illiterate and semi-illiterate populations during the early part of the twentieth century. According to Diana Miliotes, Posada's illustrations offered portraits of "sensational crimes, oddities and innovations, satirical images regarding local customs, popular song sheets and humorous commemorations of the Day of the Dead (All Souls Day) celebrations among others" (11–12). As John Lear has shown, Posada's broadsheets, while more often moralistic or satirical than overtly pro-worker, also reflected the highly charged atmosphere between labor and bosses in the period 1908 to 1913 (29–43). One of the most famous of Posada's *calaveras* is *La Catrina*, a female skull wearing an elegant upper-class woman's hat, which came to symbolize the European pretentions of Mexico's upper classes—and, by extension, the elitist outlook of Porfirio Díaz's regime. As a religious warning or memento mori, such images asserted the equalizing power that death exerts, a message that appears to presage the devastation of the coming Revolution, especially in Posada's 1910 *Calavera Revolucionaria* (Revolutionary calavera), also known as the *Calavera Oaxaqueña* (Calavera from Oaxaca), in which a large *calavera* dressed in a traditional Mexican horseman's gear or *charro* outfit threatens violence as he brandishes a machete while smaller *calaveras* flee in his wake (Miliotes 36).

After the Revolution, Posada's images, especially the *calaveras*, were rediscovered by artist Jean Charlot, whose work has been reinterpreted as a precursor of the bold styles of the muralists of the post-Revolutionary period, including José Clemente Orozco and Diego Rivera (Miliotes 4–5). Mexico began to rebuild its society, beginning with the 1917 Constitution and the founding of the Partido Nacional Revolucionario or PRN in 1929 (renamed the Revolucionario Institucional or PRI in 1946). However, despite hopes that it would protect all Mexicans and their interests, the institutionalization of its one-party system eventually led to dissatisfaction and corruption, and by the end of the twentieth century, complex factors, including internal political dissent and neoliberal policies had eroded any will to protect national interests.

Zombie-*calaveras* emerge in Mexico in Diego Velázquez Betancourt's 2013 *La noche que asolaron Tokio* (The night they destroyed Tokyo) according to Carmen Serrano ("Mapping" 462). Velázquez Betancourt uses parody and literary experimentation to contest the fears aroused by zombie horror narratives in his three-part novel. The protagonist, Andrés, a clerk who works at a store whose name bears a marked resemblance to Walmart, represents the typical alienated worker who is barely surviving. He attempts to make his life more meaningful by participating as a stagehand in a theatrical production called *La noche que los zombis asolaron Tokio*, but this experience only reinforces his alienation, since neither the script nor the actors communicate in a meaningful way. Returning to his everyday existence, Andrés finds the world around him deteriorating into an apocalyptic landscape. As he wanders across Mexico City, he comes across several solitary zombies who, significantly, are more comical than menacing:

they engage in everyday activities such as feeding birds, who eventually feed on them, and hide their fingers when they fall off so as not to frighten the living. In this respect, according to Serrano, they are more evocative of Posada's *calaveras* than traditional zombies (466). In the end, Andrés shares their fate, as he becomes a zombie after being beaten and handcuffed to a post by a madman who leaves him to die for no apparent reason. Serrano interprets Andrés's death as a reflection of Mexico's recent onslaught of violence, including the "murders of young women in Ciudad Juárez, student killings dictated by government officials, [and] assassinations brought on by drug violence" (469–70).

Before he dies, Andrés tries get a foreign woman to free him, but she abandons him to his fate. Utterly traumatized, he "dies" and becomes a zombie. Even so, he still tries to connect with others, but is unable to communicate: "Yo los llamo. Agito mi mano y los llamo. Agitar mi mano espanta a las palomas y a los zopilotes y a los tucanos que me mastican. Se me cae la mano" (276; I call to them. I wave my hand and call to them. Waving my hand scares away the doves, and the vultures and the and toucans that are chewing on me. My hand falls off.) The waving hand that falls off offers an absurd ending of the zombie narrative as horror, turning bodies into a necropolitical commentary. According to Sherryl Vint, the capitalist economic system "separates the human spirit from the human body . . . [that] lives to work (that is, produce surplus value for capital)" (178). Ironically, at the moment when Andrés's hand is separated from his body, he is trying to use it for communication rather than to work and accumulate capital.

Brazil's population has also been disillusioned with its neoliberal expansion, since it has not managed to fully realize the democratic objectives of the 1988 Constitution, and unequal political, civil, and material rights continue to plague society (Lehnen 7–9). James Holston has characterized Brazil's process of democratization and urbanization as "contradictory," in which the full rights of citizenship have not been extended to the lower classes, since restrictions on political power and education, along with institutional violence, have prevented social progress (245).

Fear of corruption and a sense of disillusionment are themes of André de Leones's novel, *Dentes negros*, published in 2011 toward the end of the mandate of Luiz Inácio da Silva (2003–2011). The novel uses the viral metaphor to comment on the "diseased" body politic. It portrays the survivors of a plague that has left all but 8 percent of the Brazilian population dead, with their mouths open and dark teeth exposed. Told mainly from the point of view of two survivors, Hugo and Renata, the story evolves as they travel to the state of Goiás, in central Brazil, in order to join Hugo's only surviving cousin. The novel takes place in a near-future Brazil whose social reality—class structure, television, restaurants, family structures, and internal migration north to south—remains unchanged from the present. Renata plainly states that even though the government has contained the virus and even developed a vaccine against the infection, she believes that everyone is still infected. In fact, she believes that it is in their bones, like radiation. Their sense of rootlessness, lack of family, distrust of the government, and fear of violence make these survivors into trauma zombies. Leones envisions a dark future, with roving bands that recall actual zombie hordes, as Hugo and Renata are attacked on their trip and later die after seeking help at an army base.[8] Leones's novel captures the idea of failed democracy and failed immunity by using several types of metaphorical zombies: those whose black teeth and open mouths appear to hunger for an unrealized future, those who resisted actual zombiedom but are left wandering, hoping to stay alive, and those who compose the roaming hordes who attack them.

Homero Aridjis's 2014 novel *Ciudad de zombis* uses zombies to address the issues of drug violence and human trafficking that have been incorporated into Mexico's body politic. In the novel, there are two types of zombies: *los vivos muertos* (the dead living), criminals who are involved in the drug trade and prostitution, and *los muertos vivientes* (the living dead), people turned into zombies by trauma. Significantly, the zombie king or drug lord of Aridjis's novel is named Carlos Bokor; a *bokor* is an evil sorcerer in Haitian voudou. Bokor presides over a city where the criminal zombies act in alternately horrific and normal ways, at times invading stores and driving away regular customers, at other times sitting quietly in buses or eating at the Trendy Restaurant, a venue designed especially for them. Jen Webb and Samuel Byrnand remind us that this kind of zombies—criminals and traffickers—tend to inhabit the same modern environments we do, living in groups and spreading to establish their own hegemony over other people and places (111–12). In Aridjis's novel, this scenario takes on an eerie familiarity with criminal zombies coexisting with the human inhabitants of the city. This uncanny sense that zombies are doubles for humans is used by Aridjis throughout the novel, illustrating Mexico's take on shifting attitudes toward the zombie figure. It is clear as the novel progresses that the second kind of zombie, the *muertos vivientes*, are victims of trauma.

The story begins in earnest when an investigative journalist living in Mexico City, Daniel Medina, learns that his ten-year-old daughter Elvira has been abducted. Her trail leads him to an auto chop shop whose catalogue lists different models of automobiles as code for underage prostitutes in the town of Misteca, a composite of Mexican cities plagued by violence and the hub of the human trafficking ring headed by Bokor.[9] Daniel's search appears to be of no interest to authorities, who cannot be bothered with crimes involving the lower strata of society, that is, victim or trauma zombies.

Significantly, this novel expressly mentions "outbreak," "contagion," and the "immunological" response. Eventually, through social media, the upper classes become aware of the presence of criminal zombies and subsequently demand a response from the government. Forced to act, the government takes steps to eradicate the infection, sending a group of plague doctors and specialists to enact protective immunological measures aimed at containing the *muertos vivos* of the drug trade. The violence escalates to assault and the burning of zombie neighborhoods, followed by the work of clean-up crews, who eventually find a large number of kidnapped young girls in Carlos Bokor's bunker, including, it seems, Daniel's daughter, Elvira. This, however, turns out to be a failed immunological response by authorities, since it does not eradicate the problem.

Although the crime appears to be solved and the criminals dead, nothing is ever as it seems in *Ciudad de los zombis*. We learn that Bokor has doubles who appear out of nowhere at different locations, and that it is even possible that the actual Bokor has escaped via a tunnel leading to the Pacific, recalling the legendary escapes of the actual drug kingpin of the Sinaloa cartel, Joaquín Guzmán Loera, El Chapo. In Aridjis's novel, Bokor sends a text informing the public that "el monstruo nunca muere" (312; the monster never dies). This emphasizes the fact that the elimination of one person has little effect on the structure of crime networks, because, as Beckman points out, "the narcotics trade serves as but the latest articulation of a long history of export commodity booms, creating unprecedented amounts of wealth" (xxix). Aridjis may also be commenting on the failed campaign begun in 2006 by President Felipe Calderón to eradicate the cartels in his home state of Michoacán, which only increased the number of deaths and caused the cartels to specialize and branch out into other activities, as demonstrated by the activities of Bokor.[10]

As the story evolves, Daniel undergoes a strange transformation, almost imperceptibly turning into a trauma zombie. From the beginning, we wonder why he chooses to stay at the Gran Hotel de Misteca, when the staff of the hotel expressly warns that "El huésped podría ser devuelto a su lugar de origen convertido en zombi en ataúd de cartón" (46; The guest can be returned to his place of origin converted into a zombie in a cardboard coffin), and indeed, after a few days, Daniel states that, "noté rasgos de zombi en mi cara: el cráneo calvo, las mejillas descarnadas, el mentón pelado, el cuello enegrecido" (95; I noted traces of zombie in my face; a bald head, sunken cheeks, a skinned chin, and a blackened neck). At first it seems merely metaphorical, but later we learn that he has been shot and taken to the morgue as a John Doe, where "me colgaron una etiqueta del cuello" (156; they hung a label from my neck). However, a friend gets him out of the refrigeration unit and takes him back to his hotel, where he appears to reanimate. Despite this recovery, he later refers to dressing his *cadáver* (179), suggesting that his zombification is complete, although he continues to act and narrate as if nothing has happened. When a man claims that a girl he has recovered is Elvira, Daniel greets her warmly and the two take a bus back to the capital. When he finally takes a good look at her, wrapped in a serape, he remarks that she is: "Una criatura extraña, mayor que Elvira, algo marchita y tan vieja que parecía sin edad" (330; A strange creature, older than Elvira, and somewhat dried out and so old as to appear to be ageless). The girl, by wearing an of traditional piece of Mexican clothing and having an ageless look about her is an emblem of the historical suffering and collective trauma of contemporary Mexico.

In Aridjis's novel, both Daniel and the girl develop their own response to corruption and suffering by becoming trauma zombies. They are the living dead, who must endure the colonizing forces of the American War on Drugs, corrupt Mexican authorities, drug violence, and criminal activity, surviving catastrophe after catastrophe. Marco Vladimir Guerrero Heredia has used the word *narcogótico* to describe the genre of *Ciudad de zombis*, noting that the bodies in the novel are used to convey anguish, not fear: "Nada se contiene, nada se oculta; el cuerpo antes humano se revela hasta los huesos, se vuelve entraña, pedazo" (315; Nothing is contained, nothing is hidden; the formerly human body is revealed down to the bones, becomes entrails, a piece). These zombies are meant to bring us back to our humanity by exposing us to the anguish of those who die but refuse to disappear in the collective memory.

Consumer Zombies: Baroque Resistance in Parody and Play

Latin American zombies epitomize another feature that derives from Andrade's weapons of choice for social critique as expressed in his "Cannibalist Manifesto": play, parody, or humor.[11] In his book *Improper Life* (2011), Timothy Campbell suggests that the principle of play through imagination and language can dismantle or interrupt discourses of power (153). To the extent that consumer zombie texts from Mexico and Brazil often mock capitalist principles and break down the actions and language of capitalist logic, they are subversive. In his 2007 essay "In Praise of Profanation," Giorgio Agamben proposes that, in order to restore meaning to actions outside of capitalist relations, we must "profane" that which is now "sacred," that is, work, obligation, and productivity (74–81).[12]

I would argue that the new perspectives on the zombie body that we have seen up to now can recalibrate our sense of humanity and the significance of embodiment and borders between binaries. Similarly, if humor disarms us, then certain zombie narratives can estrange us by satirizing our patterns of consumption or other mindless actions. Mexican and Brazilian consumer zombie narratives often use exaggeration to mock the values of efficiency and the capitalist logic of consumer transactions. Webb and Byrnand describe our consumerism (following Lacanian theory) as the search for the missing part of ourselves, creating an inner zombie of consumption (115), as if "our consciousness of self and control of self have been seized by a bottomless hunger to consume no matter what, and no matter that it does not nourish us" (118). Modernity and consumption for its own sake, along with the mindless pursuit of efficiency and productivity, contribute toward making people into zombies. Humor or "profanation," to use Agamben's term, can be used to undermine capitalism's soulless logic.

As early as 1940, Bernardo Ortiz de Montellano anticipates our current paradigm of consumer zombies in "Cinq heures sans coeur" (Five hours without heart), a story written in Spanish but inexplicably given a title in French. In the story, a race of tiny humans plans to take over the world by replacing humans of normal size. Driven by extreme efficiency to consume, they have no emotions. Each of the five parts of this satirical story represents an hour of the miniature narrator's five-hour life span, allowing readers to understand his worldview. When the narrator, A, mentions that he must meet with E, a lovely female with whom he will immediately consummate a union, he complains that the traditional rules of courting are inefficient. He then explains that the purpose of life is to produce consumer citizens: "la fábrica-escuela de los nuevos seres humanos [es un] ejército de consumidores, el 'salvation army' de nuestra economía" (128; the factory-school of these new humans [is the] "Salvation Army" of our economy). The fast-paced, unfeeling consumerism of these beings recalls the relentlessness of modern zombie hordes. Ortiz de Montellano is expressing anxiety about intellectual ennui and the replacement of the arts by material values, what Phillip Mahoney has characterized as a general horror of "figures of the multiple, zombies and crowds confront[ing] us with a kind of uncanny featurelessness" (116), of the cultural paradigm shift of urban modernity.

Dating from the 1940s, Montellano's story draws inspiration from industrial paradigms of assembly-line efficiency. In her work *Shifting Gears* (1987), Cecilia Tichi notes that early twentieth-century advertising constantly compares the human body to an engine, with food described as "Fuel for the Human Engine" (6–39), while "waste" and "inefficiency" indicated the need for product redesign (64). It is clear that these concepts are parodied in the work of Ortiz de Montellano, in his notions of speed, productivity, and consumerism, illustrating how the American concept of "Fordism"—the use of assembly lines and the language and principles of industrial production—were perceived in Mexico. While the correlation between the zombie drone worker and the assembly line is clear, these miniature humans are highly organized and even have a universal language with which they communicate through radio waves in a form that anticipates the instantaneity of the internet. These zombie-like characters are portrayed as lacking individuality or subjectivity, confirming Webb and Byrnand's analysis of zombie purposelessness, of "non-thinking consumption" performed at viral speed (117).

Manuel Becerra Acosta's 1945 story "El laboratorio de espíritus" (The laboratory of the spirits) relates zombification and Americanization in workers who give themselves over to a biopolitical process by which they are repeatedly refurbished as satisfied workers. The story consists almost entirely of a promotional text, narrated by the director of the Hamilton Institute, that promises its clients total rejuvenation, first through plastic surgery to improve their appearance, then through an intervention to revitalize them physically, and finally through a complete psychological makeover to improve their outlook. The potential client for this treatment marvels at the director's words, commenting that they sound strangely familiar: "Aún repercutían en mi interior las palabras tan persuasivas de aquel hombre, cuando desperté. . . . Han transcurrido los años—que muchas veces he creído que *he vuelto a vivir*—pero nunca me ha sido dable precisar si efectivamente fui un pensionista del 'Hamilton Institute.'" (115, emphasis added; That man's persuasive words still rang in my ear when I awoke. . . . Now many years have passed—which many *times I felt that I have lived before*—but I have never quite figured out if I was ever a patient of the "Hamilton Institute").

The fact that the name of the treatment center appears in English as the "Hamilton Institute" in a text written in Spanish inevitably associates it with the culture of the United States and its increased influence in postwar Mexico. Becerra Acosta himself spent time working as a journalist in New York (Lara Pardo 12), and it seems that this experience led to a critique of zombie-like behaviors he observed there, in which conformity and happiness seem to be artificially induced and any negative memories erased, resulting in docile workers who are easily controlled. Biotechnology combines with capitalism to produce an ironic warning about the danger of creating zombie-like citizens who lack any memory of the past and live in an eternal mindless present.

A different type of consumer zombie, more akin to the somnambulist type characteristic of Hollywood films, appears in a story by Brazilian poet Carlos Drummond de Andrade, "Flor, telefone, moça" (1951). It begins in the cemetery of John the Baptist in Rio de Janeiro, where an adolescent girl picks a flower from the grave of an unknown man and carelessly tosses it away. This deflowering reverses the usual association of violation and virginity, but the act results in punishment for the girl and the members of her family, who are unable to restore her peace of mind when the phone rings and a mysterious male voice demands the return of the flower to the grave. In Drummond's story, despite the daughter's reasoned answers and pleas, the voice calls her every day to make the same demand. Living in dread of the daily phone call, she begins to lose all interest in life. After hearing of her troubles, her father, mother, and brother attempt to help her, yet one by one, each fails to appease the voice. The young woman descends into a zombie-like trance and finally succumbs to her suffering.

Drummond's story could be interpreted as a conventional cautionary tale about the dangers of disturbing the dead and ghostly haunting, but at the end it takes a quick turn to the absurd. Since the mysterious voice is like a modern *bokor*, it uses technology, in this case a telephone instead of a hypnotic spell, to control the female zombie victim. When the father goes to the telephone company to disconnect his service, a telephone executive explains that do so, "Seria uma loucura. Então é que não se apurava mesmo nada. Hoje em dia é impossível viver sem telefone, rádio e refrigerador" (24; It would be madness. Plus, it won't help anything. Nowadays it is impossible to live without a telephone, radio, and refrigerator). In the face of these arguments, the father relents, showing that the family is not willing to disconnect itself from modern amenities even if it means losing their daughter. Her death is a price they are willing to pay in order to remain connected to modernity and its conveniences. Drummond seems to anticipate our current attachment to cellular phones, and those of us who receive and tolerate incessant calls from nonhuman digital entities may find this story more than a little unsettling. As I will show toward the end of the chapter, Karen Chacek offers resistance to this type of zombification in her critique of cellular phones and the plight of children in *La caída de los pájaros* (2014; The falling of the birds).

Political Zombies and Esposito's Immunitas

In her masterful study of archetypal modern monsters, *El monstruo como máquina de guerra* (2017), Mabel Moraña stipulates that the Latin American zombie signifies the fear of losing autonomy or individuality, an idea that is unthinkable for Western audiences but plays differently in Latin America, where a long history of oppression has impeded the adoption of the pretension of the liberal autonomous self (166). David Dalton notes that the fear of zombies is not as strong in Latin American culture, in large part because the undead have much in common with poverty-stricken local populations ("Antropofagia" 5).

Esposito's concepts of community and immunity, which I discussed in the Introduction, are useful in describing a certain form of resistance. In his book *Immunitas* (2002), he argues that what disease is to the body, social danger is to the body politic (2), and that both may be expected to generate an immune response when threatened. He further explains that, just as the body may fall victim to auto-immune diseases when immunological systems attack healthy tissue, the body politic may be harmed when its response against perceived threats, perhaps in the form of civil war or police actions, involves excessive force directed against the citizenry (15–17). Esposito's concept of immunity is of special importance for understanding resistance in texts about the living dead in Mexico and Brazil, because, as I postulate, disenfranchised groups targeted by the body politic's immune response may ultimately be forced to take counter-immunological measures based on the baroque ethos and their historical marginality as a form of resistance to state violence and coercion. In other words, as David Dalton has suggested, subalterns may resist and even subvert *immunitas*, undermining state power and creating their own form of resistance ("Antropofagia" 5). As mentioned earlier, Esposito proposes a distinct kind of "immune response," one that is based on tolerance rather than rejection (169) and that offers a respect for individuality and dialogue as an alternative to necropolitics. I feel that the literary representations of zombies in Mexico and Brazil show that the counter-immunological power of resistance has inoculated the population against state violence through the application of the baroque ethos, allowing them to resist state power.

In the 1930s, both Mexico and Brazil consolidated their modern political structures, and although they followed different ideologies, both became models of success for economic development in the decades that followed. As Douglas Graham summarizes, during this period, "Brazil experienced more impressive industrial growth, but Mexico underwent more significant political and institutional reforms" (17), and because of its revolutionary experience, Mexico had an inclusionary attitude toward labor and rural sectors while more exclusionary attitudes prevailed in Brazil. Despite its unmatched economic expansion between 1930 and 1960, Brazil experienced less stable growth, and eventually the impact of social unrest combined with inflation and economic pressures led its technocratic military leaders to seize power in the 1964 coup. Paradoxically, though they were vehemently antisocialist, Brazil's military regime encouraged the growth of public enterprises, while the civilian socialist ideology of Mexico's PRI government offered more support to local private investors (Graham 34). The bottom line is that both governments adopted some form of economic protectionism and the political fortunes of both were highly linked to economic well-being. This period also corresponds to the Cold War, when institutional powers were determined to quell internal dissent, as armies and police grew in power and increasingly focused on protecting the state against communism (Wiarda and Kline 34).

Esposito offers an example of quarantine as a possible defensive move by the state against a viral attack, in which the body politic attempts to seal itself off to preserve its integrity and prevent further "infection" (*Immunitas* 123). One zombie tale that evokes this attempt at quarantine is Bernardo Esquinca's "La otra noche de Tlatelolco" (2015; The other night of Tlatelolco), a story that reinterprets a specific event of Mexican history: the massacre of student protesters on October 2, 1968, just days before the opening of the Olympic Games in Mexico City. This attempt at political quarantine recalls how students in the demonstrations were corralled and contained within the Plaza and its surrounding streets to prevent their escape from government troops, who subsequently massacred them with no explanation or acknowledgment by the government of Gustavo Díaz Ordaz. The names and number of students killed still remain unknown. In Esquinca's story, following the massacre, several students rise from the dead—become zombies—attacking the police and turning against their oppressors. When one of the students gunned down in the massacre becomes a zombie, he bites a policeman, who then spreads the infection to hospital workers. Esquinca's story implies that the zombie dissent will continue to spread—as a form of counter-immunological reaction—undermining the government's legitimacy and power. The historical 1968 event also recalls the 2014 disappearance of forty-three students in Ayotzinapa in the state of Guerrero, another crime that has proven impossible to solve in the face of government corruption. The apocalyptic sentiment of this zombie story underscores Mexico's continued frustration with unanswered violence and the corrosion of state institutions.

The year 1968 also marks the declaration of the Brazilian military government's draconian Institutional Act No. 5, which suspended civil rights and institutionalized torture and censorship. Known as the "coup within the coup," it was a victory for the military hardliners and led to disappearances and the terrorization of the civilian population.

A number of leftist groups adopted armed resistance and operated as small cells that were eventually located and successfully "contained" and "neutralized" by the Brazilian military police by 1974.[13] Although outwardly successful, these measures had the long-term effect of undermining the regime's hold on power.

Esposito explains that one of the biopolitical mandates of the state is to control the spread of infection: "The first step is to isolate places where infectious germs may develop more easily due to the storage of bodies, whether dead or alive: ports, prisons, factories, hospitals, cemeteries" (139). In Brazilian Érico Veríssimo's 1971 novel *Incidente em Antares* (Incident in Antares), this separation of the living and the dead is violated when seven bodies are left unburied at the cemetery. Clearly aimed at critiquing the military regime and its excesses via the testimony of the undead within the broader history of southern Brazil, *Incidente em Antares* was a bestseller in its time, going through several printings in 1971 alone.

Broadly panoramic, the novel offers a portrait of Antares, a fictional town in Rio Grande do Sul, Veríssimo's home state, through an examination of the lives of its two most important families from the mid-nineteenth century to 1963, a year before the military took power.[14] The clearest reference to the military regime in the novel is a date, Friday, December 13, when the living dead appear, marking the exact date of the 1968 takeover by hardliners (Pellegrini 111n30). The anxiety surrounding torture and the disappeared is a clear referent in Verissimo's novel, especially in the figure of the political activist João Paz, whose death at the hands of the police is attributed to heart failure by a complicit local doctor. It is important to note that Veríssimo's description of torture methods used by the military regime is deeply embedded in the 485-page novel of fantastic literature, explaining, in part, how the work managed to pass through censorship at the height of the dictatorship during the 1968–1973 period.[15]

The undead do not appear until midway through the novel, when a general strike by some four hundred workers brings the city of Antares to a halt. Zombies are generally associated with the working classes, but here it is Quitéria Campolargo—the city's grand dame— who is first to become a zombie, followed by the workers. This again shows that the living dead in Brazil represent all social classes and even retain their sense of class identity after equalization through death. The living dead in Veríssimo's novel are not feared but instead initiate a campaign to inoculate the people in their struggle against power.

The zombies from Antares are not the usual shuffling masses, like the somnambulists of Haitian tradition, but are instead truth-telling members of a cross-section of the town's society whose comments focus on outdated social structures, hypocrisy, and corruption. In this sense, Veríssimo recalls the Portuguese tradition of the *auto* or morality play that was adapted by the early-modern playwright Gil Vicente (1465–1536) to produce festival plays that used humor and character types to convey their messages of morality and proper behavior. Vicente's 1517 *Auto da barca do inferno* (Auto of the boat of hell), for example, involves a group of deceased characters similar to those in Veríssimo's novel, including a nobleman, a fool, a shoemaker, a procuress, a monk, a lawyer, and a hanged man, among others. However, Vicente's Renaissance characters are souls making their case before the devil and an angel, while Veríssimo's protagonists are actual corpses, zombies who return to the town's square, a powerful place symbolizing Esposito's *communitas*, that is, the place where the state and community settle accounts and perform mutual obligations (*Communitas* 5). It is important to recall that Brazil was already living under dictatorship when this novel was published. Although they project an abject, death-like presence through their ghastly appearance and odor, the zombies are not viewed by the townspeople as Others, whose extermination was justified by the government's "state of exception" (9), but instead as initiators of a much-needed political debate.

As the seven decaying zombies walk into town, they are surrounded by flies buzzing around their heads, signaling the start of the swarm counter-immunological response.[16] Later, the appearance of a different type of swarm—rats—intensifies the response. The rats appear after the zombie of João Paz describes in graphic detail the torture he endured. Additionally, the zombies reveal that the police and mayor are all criminals, that the prosecutor and the previously revered teacher have unsavory sexual appetites, and that the town doctor has acted in collusion with the mayor and police chief. It is at this point that rats begin to swarm the city, unafraid to stand up to authority: "Alguns revelavam uma audácia e uma agressividade até então desconhecida dos antarenses que, assustados, os viam entrar nos guarda-comidas, fossar nas latas de lixo, subir nas camas, enfrentando sem temor e às vezes sem recuar, gritos humanos, vassouradas e até o assalto de cães e gatos mobilizados para combatê-los" (377; Some [rats] revealed an audacity and an aggressiveness never before seen by the inhabitants of Antares, who, frightened, saw them go into their pantries, dig around in their garbage cans, get into their beds, facing them without fear and undeterred by human screams, brooms, or attacks by dogs and cats mobilized to fight them). The swarm of rats represents the people, whose political demands, like those of the dead, must be met. The regime's use of secrecy and force leads to corruption, triggering its autoimmune response. Although the authorities appear to make concessions, they continue to maintain power through illegitimate means. While offering the striking workers a decent wage, the town leaders secretly initiate a publicity campaign to erase the event from memory (known as Operação borracha [Operation rubber]), claiming that the zombie incident was a fabrication at best and a lie at worst (i.e., what would now be called "fake news"). Only a few stubborn eyewitnesses, among them the local photographer, the progressive priest Pedro Paulo, and a university sociologist, maintain the "truth" of the zombie invasion, resisting the narrative of the more powerful. Just four years after *Incidente em Antares* was published, historical events in Brazil confirm that the public was fed up with the excesses committed by the regime. One example is the public reaction to the torture and murder of journalist Vladimir Herzog in 1975 in São Paulo, which brought thousands of protesters to the Praça da Sé in one of the first open manifestations since the takeover in 1968 by hardliners and their crackdown on political dissent through music, theater, journalism, and books (Piletti 182).

Lygia Fagundes Telles's 1977 story "Seminário dos ratos" (The rats' seminar) also employs the motif of the swarm in a satire denouncing political violence and repression.[17] In Telles's story, while hosting a summit of the hemisphere's most important military regimes and their American advisors, Brazilian officials are embarrassed by a rat problem first discovered in the favelas, where it spreads like a plague. The rats multiply to the point that they are found everywhere, even in the mansion where the governmental meeting is taking place. Telles's rats represent the sense of horror that the military leaders have toward their own citizens, who appear to them as a relentless swarm of vermin marauders. As Dalton notes, Latin American zombies (and other abject beings) do not always represent a plague against which humanity must fight, but instead can act as a force that "vaccinates the region from the imperialistic tendencies—both internal and external—that still lie dormant within its population" (13). In Telles's narrative, the rats/zombies inoculate the society with resistance against the imperial powers symbolized by the American presence at the summit meeting. By 1978, the Brazilian military leaders began a process of political transition and a loosening of control, phased in as *distensão* (political easing up), followed by the *abertura* (political opening) in 1980 and the eventual transition to civilian rule in 1985.

Another version of the immunological response is exemplified in Karen Chacek's 2014 short novel of the fantastic, *La caída de los pájaros* (The falling of the birds), which illustrates both the body politic's immunological reaction to a perceived threat and a counter-immunological response by those that come under attack. At the beginning of the story, all birds suddenly and inexplicably die, traumatizing the children who witness the event to the extent that they fall into a coma and have to be interned in a hospital. After several months, their parents stop visiting them at the hospital when a video feed is provided showing the continued care of the children. The children eventually recover and desperately try to communicate their presence to family members, but they have become invisible to them. The narrator and protagonist Violeta comments, "La ciudad está llena de personas convencidas de que los muebles y adornos de sus casas se mueven de lugar, misteriosamente, al menos una vez por semana" (55; The city is full of people convinced that their home's furniture and knickknacks mysteriously change place, at least once a week). The parents fail to realize that it is the children who are moving the furniture and writing messages on the city's walls.

Subsequently, Violeta, herself traumatized by a similar event as a child, begins to hear a child's voice in her head and learns of the children's true situation through a comic-book artist.[18] The artist reveals that, with the economy booming, the government now considers the children to be a threat, because they distract their parents from productive work. In order to maintain levels of productivity among workers, government agents engage the immunological apparatus, hypnotizing parents into believing that their children remain unconscious. Violeta subsequently gives the children crayons and mails their letters and artwork to their parents in a successful effort to convince them that their children are actually awake, counteracting the hypnosis.

The story functions as commentary on cyber-addictions and how new technologies have caused adults to ignore their families and shun "nonprofitable" moments of leisure, enjoyment, and intimacy. Chacek uses the visual arts—as shown by the prominent role played by the comic-book artist and the children's graffiti—to break the spell of the government agents bent on protecting economic interests and ensuring continual economic growth. Thus, the children are able to activate a counter-immunological response because they are outside of capitalism and connected to the ludic aspects of life. Because play allows for the dismantling or interruption of power— as in Agamben's concept of profanation—Chacek uses children and their natural predilections in her novel to break the biopolitical hold that induces adults to work constantly. Finally, it is notable that the children begin as somnambulists or quasi-zombies, but later express resistance through illegal graffiti, while the parents are turned into zombies by government actions. It is the children who are eventually able to counter-immunize their parents against the virus of unending productivity.

Conclusion

In this chapter on the undead in Mexico and Brazil, we have examined the zombie and its various iterations mainly through Esposito's theories of immunity and the body politic. Trauma zombies emerge when populations with no immunological resistance are faced with sudden economic and political change. Consumer zombies, meanwhile, appear as representatives of the anthropophagic or parodic aspect of the genre, where they are used to critique economic systems through parody and profanation. Finally, the most original iteration of the zombie are what I have called resistant zombies, zombies that enact counter-immunological responses designed to protect themselves when targeted by the state's immune or autoimmune response. As expressions of the baroque ethos, these resistant zombies attempt to stave off the contagion of violence perpetrated by their bodies politic, gesturing "towards awakening and political action" (Serrano 471) in their continuing role as resistant Other.

CHAPTER 4

Vampires
Immunity and Resistance

Although this chapter is structured around Esposito's concept of immunity, I wish to emphasize that Echeverría's baroque ethos and Andrade's *antropofagia* continue to be relevant, since successful immune responses to attacks by vampires, these powerful monsters of the undead in both Mexican and Brazilian speculative fiction, must incorporate baroque resistance and anthropophagic adaptation.

As a prelude to contemporary vampire narratives in which vampiric elites are portrayed as predators, it seems appropriate to recall Machado de Assis's 1870 story "A vida eterna" (Eternal life), which features an elite secret society whose members' power derives from eating human flesh. This cannibalistic perpetuation of power is a reference to secret alliances that have traditionally fed on the general population, showing that the decolonial view of elites as vampiric predators is not entirely new in Brazil.[1]

In *Bíos: Biopolitics and Philosophy* (2008), Esposito examines several vampires or vampire-like figures, including Dr. Jekyll, Dorian Gray, and Stoker's Dracula, all of whom personify late nineteenth-century anxieties about the contamination of the blood, degeneration, and contagious diseases (124–26). For Esposito, the unclassifiable body of the monster (half human, half animal) connotes a breakdown of the typological categorizations, which was equated with degeneration in the late nineteenth century (119). Dracula becomes a parasite that embodies the principle of contamination because he "lives on the blood of others and reproduces that way" (126). Esposito explains that, when a community is "forced to introject the negative modality of its opposite" (52)—here the monster or the poison—in order to develop antigens for an immunological response, "the neutralization of conflict does not completely provide for its elimination" (62–63), that is, the monster/poison does not completely disappear. Perhaps for this reason, as a representative of the undead, the vampire is particularly effective as a revenant monster that represents recurrent and unresolved historical issues.

In this chapter on vampires I will begin with stories that show how Esposito's concept of *immunitas* captures humanity's role in containing the threat of foreign vampires.[2] These are Alejandro Cuevas's "El vampiro" (1911; The vampire) and Carlos Fuentes's "Tlactocatzine en el jardín de Flandes" (1954; Tlactocatzine in the garden of Flanders), in which foreign vampires invade urban homes to feed on the living, provoking a defensive immunological response. A distinct kind of resistance can be found in Amparo Dávila's "El huésped" (1959; The guest) and Lúcio Cardoso's *Crônica da casa assassinada* (1959; Chronicle of a murdered house), in which women must struggle against mysterious vampiric forces in the tropical gothic mansion in order to resist and achieve liberation. I follow up on these two narratives of female resistance with a story from 1990, Gabriela Rábago Palafox's "Primera Comunión" (First communion), which recasts resistance to the patriarchal Catholic Church through lesbian sexuality.

A third group of stories explores how alien or biologically engineered vampires actually work on behalf of subaltern groups, conferring Esposito's *bíos* or political legitimacy on humans and themselves. Among the initiators of this paradigm shift are Gerson Lodi-Ribeiro, in his cycle of the alien vampire Dentes Compridos (Long Teeth) of *O vampiro da Nova Holanda* (1998) and Gabriel Trujillo Muñoz, in *Espantapájaros* (1999), in which a vampire becomes an ally of those marginalized by society, using its power to transform the body politic, an agent of what I have called the counter-immunological response. Finally, I present a set of stories about resistant humans who are neither predator nor protector, but use the power of vampires to gain a wider sense of self and community, renewing their commitment to the *communitas*. This kind of interaction suggests a different understanding of the body and personhood as what Esposito calls the "impersonal" or the sacred (Campbell 78),[3] bridging the divisions of social class and the dichotomy of human versus vampire/monster. To illustrate this fourth type, I have chosen José Luis Zárate's *La ruta del hielo y la sal* (1998), André Vianco's *Os sete* (1999) and Giulia Moon's *Kaori: Perfume de vampira* (2009). I would emphasize that this final type, rather than using concepts of blending and hybridization and the language of *mestizaje/mestiçagem* to legitimize national identity, involves characters who, in an anthropophagic gesture, absorb the bite of the vampire rather than being transformed by it, using blood as an antigen to reestablish bonds of community. This approach demythifies the narratives of the "cosmic race" or the "fable of the three races" associated with *mestizaje/mestiçagem*. As Jacob C. Brown has argued in his study on Brazilian vampires, *mestiçagem*, like the vampire itself, is one of Brazil's "persistently undying and undead" national myths (21). Since I believe that discourses of hybridity can be used to hide historical violence and suppress political and social community, I emphasize these characters' resistance to hybridity and their engagement in activities that help to break down the separations fomented by necropolitical policies and the insularity of personal gain.

History of the Vampire Figure

In order to understand the reformulation of the vampire in Mexico and Brazil, it is necessary to understand its European origins and conventions. The vampire has deep Old World roots, traceable to Romanian nobleman Vlad Tepes, also called Vlad the Impaler (1431–1476), and to a series of wars, plagues, and exhumations that took place in Eastern Europe during the Hapsburg regime.[4] All of these elements contributed to the eighteenth- and nineteenth-century versions of the vampire familiar to us today. In his study *The Living Dead: A Study of the Vampire in Romantic Literature* (1981), James B. Twitchell examines several canonical nineteenth-century authors and their use of the vampire figure, including Coleridge, Shelley, Keats, Poe, Emily Brontë, Oscar Wilde, and Henry James. In his analysis of Keats's 1820 poem "Lamia," which takes its name from the vampire-like snake / woman figure of Greek mythology, Twitchell argues that the vampire is one of the most enduring mythic archetypes inherited from Romanticism, a supernatural phenomenon used as a metaphor for artistic creation or fatal romantic entanglements. He also hints at the existence of what I am calling "metaphorical vampires"—humans that corrupt and absorb the psychological energy of others, spreading ill will like an infection and destroying the lives of others with malice or indifference (173–76). A similar analysis can be found in Heide Crawford's explanation of the folkloric vampire in her 2016 study *The Origins of the Literary Vampire*, where it is described as a ghostly presence that made its way from Eastern Europe through Germany to England, feeding on the life force of humans (xiv).

Many vampire narratives dating from the late eighteenth and early nineteenth centuries are part of the gothic strand of Romanticism, while texts appearing later in the nineteenth century are more properly classified as belonging to the gothic revival, which spawned well-known classics such as Robert Louis Stevenson's *Strange Case of Dr. Jekyll and Mr. Hyde* (1886), Oscar Wilde's *The Picture of Dorian Gray* (1890), and Bram Stoker's *Dracula* (1897). Although the first two of these are usually not considered to be vampire texts, Judith Halberstam classifies them as such, observing that Hyde's threat to "swallow up what remains of Jekyll, is specifically vampiristic" (76), while Dorian Gray's murder of his portraitist, Basil Hallward, can be compared to a vampire attack, since Gray cuts his throat and allows him to bleed to death (77). For Halberstam, this doubling or secret self represents the monstrous body that must be disciplined and brought under control, a warning to Victorian readers against physical degeneration and psychological and moral devolution (72).

Since the 1970s, critics have focused on the vampire in Bram Stoker's *Dracula* (1897), which is associated with ambiguous sexuality, gender inversion, and homoeroticism, to demonstrate the centrality of the body in vampire tales.[5] Stoker's vampire has also been interpreted in more sociological and political ways, as a figure of racial fear, monopoly capitalism, and colonial anxiety, all of which have applications to the reality of the Latin American vampire. Judith Halberstam associates Stoker's Dracula with racist characterizations through his love for gold and his wanderlust (89), while Franco Moretti considers the vampire to be a "totalizing monster" (68) whose ability to reproduce quickly and efficiently represents the unlimited expansion of monopoly capitalism (74). In his view, Stoker's Dracula represents a kind of capitalism unaffected by ideology or religion, whose code violates the established cultural ethics of Victorian England. Above all, Dracula is considered to be a foreign threat, provoking horror in the English and their allies, who set out to kill him.

Conversely, Stephen Arata construes Dracula to be an avenging agent of colonized subjects, an example of reverse colonization. Arata notes that the vampire is stronger and more prolific than his English opponents, making him a threat to England's racial "body politic," which must stave off any menace to its purity. In this interpretation, Dracula embodies the British Empire's fears of miscegenation as it begins to fade toward the end of the nineteenth century. The fear of racially tainted blood, for example, is raised by Dracula's feeding on English women. With the gothic revival, we begin to see the vampire's ties to the monsters of capitalism and, especially, colonialism in its representations of the racial Other. Barbara Creed's examination of female sexuality and the female vampire is also helpful in examining several neo-gothic works from Latin America, especially in the confines of the gothic mansion and the Catholic Church.

The study of Latin American gothic and horror literature is now a burgeoning field. Gabriel Andrés Eljaiek-Rodríguez argues, for example, that the gothic can absorb and invert the poles of Latin American Otherness and European civilization through an anthropophagic dialogue that alternately appropriates, criticizes, and mocks the conventions of the horror genre (13). Eljaiek-Rodríguez's 2012 study "Selva de Fantasmas" examines gothic novels and films that use Latin American settings to critique the traditional monstrosity of the colonial Other (12).⁶ Among the book-length studies that examine vampires are Persephone Braham's *From Amazons to Zombies* (2015), Justin D. Edwards and Sandra Vasconcelos's *Tropical Gothic in Literature and Culture: The Americas* (2016), Mabel Moraña's *El monstruo como máquina de guerra* (2017), and Sandra Casanova-Vizcaíno and Inés Ordiz's *Latin American Gothic in Literature and Culture* (2018), all of which give considerable coverage to this monster in Latin American literature and culture (324–32). In film studies, Gustavo Subero's *Gender and Sexuality in Latin American Horror Cinema* (2016) dedicates an entire chapter on gender ambiguity and queering to Brazil's most famous cinematic vampire, Zé do Caixão (Coffin Joe), while Carmen Serrano's 2016 essay "Revamping Dracula on the Mexican Silver Screen: Fernando Méndez's *El vampiro*" and Persephone Braham's section on Guillermo del Toro's vampire film *Cronos* (1993) in *From Amazons to Zombies* (2015) address nationalism and the vampire in Mexico. Studies such as Dorothea Fischer-Hornung and Monika Mueller's *Vampires and Zombies: Transcultural Migrations and Transnational Interpretations* examine monsters in a global context, while Cathy Jrade's study *Delmira Agustini, Sexual Seduction, and Vampiric Conquest* (2012) focuses on canonical literature and the metaphorical vampire in the work of this Uruguayan poet.

One of the more focused studies of the Latin American vampire is Elton Honores's *Los que moran en las sombras: Asedios al vampiro en la narrativa peruana* (2010; Those who live in the shadows: Approaches to the vampire in Peruvian narrative). In his introduction, Honores speculates that the vampire is a *cuerpo vacío* (empty body) that can be filled by any number of psychological or cultural fears. I would contend, on the contrary, that the vampire is usually tied to a very specific time and space, and that, unlike the zombie with its anonymous mass identity, the vampire has a more individual identity in its interactions with humans. We will see that, in Mexico and Brazil, the parasitic colonial vampire and the counter-immunological vampire have distinct roles and characteristics. My own 2003 "Vampires, Werewolves and Strong Women" and Jacob C. Brown's 2019 "Undying (and Undead) Modern National Myths," for example, illustrate the Brazilian vampire as avenger and as exploiter, respectively.

Vampires in Domestic Space: Immunological Resistance

My analysis of vampires in domestic spaces begins with a story from Mexico, Alejandro Cuevas's "El vampiro" (1911; The vampire, which takes place within the confines of a home, where a vampiric invader provokes an immunological response and resistance in a child, who employs imaginative and psychological defenses. The story was written at the start of the Mexican Revolution, when nationalist sentiment was especially high, which is reflected in the fact that the antagonist is a foreigner. In the story, a man narrates a memory from his childhood. He and his father live together in an old ramshackle house after his mother's death. His usual escape from tedium consists of going onto the roof where he watches spiders fight each other or drain their victims of blood. The only visitor to the house is an elderly Italian gentleman with a beard and long teeth, a friend of his father's. The visitor provokes disgust in the boy, so he hides when the man is there, and on one occasion, the father warns the man never to go near his son. One night the boy has a nightmare that his father is being killed by a giant spider who resembles the man. When he goes downstairs, the boy finds that his father has committed suicide, leaving a note explaining that his death is the only way to protect the last of their money from the visitor, who turns out to be a moneylender.

In their introduction to "El vampiro," the editors of *Cuento fantástico mexicano: Siglo XIX* (2005), Fernando Tola de Habich and Ángel Muñoz Fernández comment on the merits of this story, especially in its effective use of setting and the overlapping images of spiders, first as part of the narrator's childhood pastime, then as the stuff of nightmares, to create an effective atmosphere for the tragic suicide (290). They also comment on the description of the *viejo-araña-vampiro* (old man-spider-vampire), yet do not pursue the implications of the figure. The description fits both the old or *ancien régime* of the parasitic upper classes and the moneylender as a foreigner or outsider, both of which provoke the boy's immunological response to protect and close himself off from this interloper.

In *Skin Shows: Gothic Horror and the Technology of Monsters* (1995), Judith Halberstam explores the discourse of anthropology on criminality and degeneracy toward the end of the nineteenth century. For example, she notes (93) that in Stoker's descriptions of Dracula, he combines racial stereotypes and criminal types proposed by Max Nordau and Cesare Lombroso (93). In Cuevas's story, the description of the Italian moneylender fits within this racist paradigm:

> Paréceme oír aún la cascada risilla y ver la figura repulsiva de aquel extranjero; su cabeza gruesa y hundida entre sus agudos hombros levantados, su cráneo de ancha frente deprimida cubierta por una montera parda, su rostro de mandíbula recia y saliente, su nariz en forma de pico de ave de rapiña sobre la boca de labios delgados y torcida por un guiño de socarrona malicia que dejaba de asomar la extremidad de un largo colmillo. (294–95)

> *It seemed as if I could still hear the harsh chuckle and see the repulsive figure of that foreigner; his large head sunk between his hunched shoulders, his skull with its wide sunken forehead covered by a brown mound, his face with a jutting jaw, his nose in the shape of the break of a bird of prey, located over his thin-lipped mouth, twisted into a squint of cunning malice that allowed a view of the tip of a long canine tooth.*

Although the Italian is not a literal vampire, his description recalls that of Stoker's Dracula with his arched nostrils, cruel mouth, and sharp teeth. Cuevas's description borders on what Esposito characterizes as the racist language of degeneration that promotes fear and separates those who are described through metaphors of "zoopolitics, of rats, insects, lice" (*Bíos* 117), a list to which might be added the spiders in Cuevas's story.

What is interesting about "El vampiro" is that it is a memory told by the narrator, a projection of his fears and loss. All of the disturbing aspects of the child's life—the spiders, the decadent house, the foreign man, and the loss of his mother—are combined into an "overdetermined" gothic monster, thus illustrating Judith Halberstam's idea of the process by which "monsters transform fragments of otherness into one body" (*Skin Shows* 92). The overlapping imagery for which the story has been praised allows us as readers to understand the construction of the gothic monster and the process by which it comes into being. Halberstam confirms that the gothic monster—especially the vampire—calls attention to the "plasticity or constructed nature of the monster" (95), as illustrated by the layers of the *viejo-araña-vampiro* of the story. The overdetermined body of the monster illustrates the workings of the immunological metaphor as a means through which the boy demonizes the moneylender in order to survive the trauma of his father's suicide. He is able to blame the outsider, performing what Esposito calls the "double enclosure of the body' (*Bíos* 141–42)" that, as Esposito notes, allows the monster to become to be "enclosed" and eliminated from the domestic body politic.[7]

Another example of the threatening foreign vampire is Carlos Fuentes's "Tlactocatzine, del jardín de Flandes" (1954; Tlactocatzine, from the garden of Flanders), which portrays a female vampire who returns from the past to haunt the present. Fuentes's story takes the form of diary entries kept by the young man who has purchased a historic home abandoned by a family leaving for Europe. He finds the house to be lovely as maintained by a pair of servants, but he is puzzled by the mysterious comings and goings of an elderly woman who appears in the evenings in the garden. He does not understand why he feels he cannot leave the house, nor why the woman seems to have no eyes. After receiving a note with the word *Tlactocatzine*—the Naua word for "emperor" (García Gutiérrez 45)—written in a spidery hand, he ventures into the garden and encounters the woman—smelling of death—who takes his hands and whispers to him, calling him Max. At this point he sees a coat of arms on a nearby door plaque that reads "Charlotte, Kaiserin von Mexiko" (45; Carlota, Empress of Mexico), and realizes that she is a vampire who has imprisoned him as her deathly lover for eternity. She is none other than the vampiristic revenant of the nineteenth-century Belgian princess Charlotte, or Carlota, as she was known in Mexico, the wife of the Austrian prince Maximilian, Emperor of Mexico (Gutiérrez Mouat 303). Placed in power by Napoleon III, Maximilian reigned between 1864 and 1867, when he was executed by Juárez's troops as they reclaimed Mexico from French rule. This event cost Carlota her sanity, which she never recovered. In the story, Carlota mistakes the young man for her dead husband and makes him her prisoner, unable to escape her vampiric presence.

Noting that Carlota is not a ghost, Gutiérrez Mouat cites her "spectral hunger for the body" (303), similar to that of a vampire who feeds off the living. In his 1981 essay on Fuentes, "La máscara y la transparencia" (The mask and transparency), Octavio Paz confirms Fuentes's obsession with this female vampiric archetype: "Es el antiguo vampiro, la bruja, la serpiente blanca de los cuentos chinos: la señora de las pasiones sombrías, la desterrada" (11; It is the ancient vampire, the witch, the White snake of Chinese stories: the woman of dark passions, the exiled). The male historian is condemned to a living death because, unlike the child in Cuevas's story, he fails to resist the threat from the past. As Genaro Pérez comments, Fuentes's protagonist "como las típicas víctimas del vampiro clásico, ni vivas ni muertas sino condenadas a existir en otro mundo infernal aparte, ya no podrá volver a la vida normal" (18; Like the typical victims of the classic vampire, neither dead nor alive, but condemned to live in a separate infernal world, he can never return to a normal life). Thus, the foreign vampire raises the issues of immunological response and fears of contagion in Mexico, since external vampires and their European influence must be contained before they prevent the country from realizing its authentic national destiny.

Part of the immunological response provoked by the vampire is related to what Rebecca Janzen denotes as "bad blood" in her biopolitical analysis of Mexican literature, especially Juan Rulfo's classic novel *Pedro Páramo* (1955). She explains that the concept of bad or impure blood, inherited from sixteenth-century Spain, applies to *Pedro Páramo*, the eponymous landowner and protagonist who imposes his will in rural Mexico after the Revolution (54). Following the idea that Pedro Páramo embodies Agamben's "sovereign power" in meting out decisions of life and death, Janzen draws parallels with Rulfo's novel in which the corruption of power (bad blood) affects Mexico's rural population years after the Revolution was fought:

> This notion of bad blood sheds a critical light on the consolidation and use of power in Mexico during the 1950. Presidents Miguel Alemán

(1946–1952) and Adolfo Ruiz Cortines (1952–1958) strengthened state power in many ways, in relation to the bureaucracy, the *Patria*, by using agrarian reform, the army, the unions and also the Catholic Church and fortified a patriarchal family structure. (57)

Janzen illustrates how Pedro Páramo spreads his bad blood to the more vulnerable through bribery and violence: he buys soldiers' loyalty with food, and he kills the local priest's brother and rapes his sister in order to control the church, thereby reducing everyone in his sphere of influence to *zoê* or killable life (57–71).

The Vampire, Female Defiance, and the Tropical Gothic

In a microcosm of sovereign power, the gothic patriarchal mansion transmits its bad blood through a vampiric presence that seeks to contain and control human behavior. When faced with human resistance, the house goes into immunological overdrive and often tries to kill those turn against it, as in Fuentes's "Tlactocatzine y el jardín de Flandes." A variation of immunological resistance is apparent in both in Amparo Dávila's "El huésped" (1959) and in Lúcio Cardoso's novel *Crônica da casa assassinada* (1959). In these two narratives, the house represents the vampiric forces of patriarchal privilege that must be resisted by women. While aristocratic privilege and physical violence connote the vampiric presence in "El huésped," this role is performed in *Crônica da casa assassinada* by the decadent patriarchal mansion. The breakdown of categories among human, animal, and plant life indicates the profound cultural threat to patriarchal power represented by women and female sexuality, which brings on an auto-immunological crisis as the house is opposed by female resistance.

Amparo Dávila evokes the vampire in her 1959 story "El huésped" (The guest), which is narrated in the first person by an unhappily married woman whose husband is rarely at home. In an isolated country house, a typical gothic setting, the wife spends her days with her two children and her maid, Guadalupe, who also has a young son.

When her husband brings home an unwanted male guest, the wife's situation takes a turn for the worse. The guest is placed in a windowless and dank corner room, where he sleeps during much of the day. When he does appear, his eyes rarely blink and he appears to stare right through other people. Fearing the worst, the wife becomes unwilling to leave any of the members of the household alone, especially after she finds the guest beating the maid's child. When the wife informs her husband of this incident, he dismisses her fears, saying, "Cada día estás más histérica, es realmente doloroso y deprimente contemplarte así . . . te he explicado mil veces que es un ser inofensivo" (22; With each passing day you are more hysterical, it's really painful and depressing to see you like this . . . I have explained to you a thousand times that he is harmless). When the husband leaves on a three-week business trip, the wife and the maid close the ventilation duct, lock the door, and nail up boards to bar the door of the guest's room, listening as he screams and beats at the door, but finally perishes. The story ends when the husband returns and calmly listens to the wife's narration of the unfortunate event. After that, the matter is mentioned no further as the story ends.

The conventional interpretation of this story centers on the tension between the husband's unconcern and the wife's feeling that the guest is a supernatural presence. In 1977, Ross Larson comments on Dávila's effective strategy of concealing the identity of the guest: "This is not the relatively simple terror of the ghost story, but that deeper anguish experienced when one feels helpless and trapped, defenseless against indefinable menace" (67). Indeed, it is the physicality or embodiment of the vampire-like guest that makes the story distinctly uncanny. The largely nocturnal habits of the guest, who prefers to live in a damp coffin-like room, feeding off the energy of others, suggest a vampiric presence. The story could also be interpreted as an allegory of domestic abuse, in which the husband's psychological domination—typical of classic gothic tales such as *Rebecca* and *Gaslight*—turns the wife into a subservient victim, powerless to fight back. The husband's denial of any threats and his insistence that she learn to live with the guest confirms this reading, as does the wife's blocked desire to escape: "Pero no tenía dinero y los medios de comunicación eran difíciles. Sin amigos ni parientes a quienes recurrir, me sentía tan sola como un huérfano" (22; But with without money and any way to communicate to the outside world, without friends or relatives to call on for help, I felt as alone as an orphan). Thus she alternates between a truce and state of siege that characterizes psychological abuse, while the maid's child experiences actual physical abuse. The issue of race is implied here as well, since the maid's name, Guadalupe, lends itself to an indigenista reading. In Mexico, the Virgin of Guadalupe is revered as the indigenous incarnation of the Virgin Mary.

The idea of a poltergeist-type presence that terrorizes women recalls Shirley Jackson's gothic novel *The Haunting of Hill House* (1959) and its film version, *The Haunting* (1966), in which an apparently malevolent force is heard but never seen. Patricia White convincingly argues that the infernal knocking in Jackson's novel that is heard only by the women and not by the men can be interpreted as forbidden lesbian desire in patriarchal culture, a clear subtext of the film (210–11). In both *The Haunting* and in Dávila's story, the male perspective does not allow for the perception of the threat, since it is outside their experience, or perhaps because the violence against women is so habitual as to be invisible in patriarchal society. In Dávila's story, it is not so much female desire that threatens the patriarchal order, but rather the women's struggle for autonomy, affirmation, or recognition of the threats against them as social and racial Others.

It is only the alliance between the two women that finally dismantles patriarchal authority and the psychological vampirism that threatens to prey upon them, along with the physical violence against the maid's child. While this element does not appear to affect the woman's marriage in "El huésped," it does solidify the bond between her and her maid, confirming their own sense of reality and agency and allowing them to avoid the gaslight moment of female isolation and insanity within the gothic house. The women in "El huésped" are able to defeat the vampire and wall off his incursion into their reality, resisting subjugation (Corral Rodríguez 220–21), constructing their own immunological paradigm. Their bond—which transcends race and social class—is the first step in containing the bad blood or necropolitical force of the patriarchal vampire.

Strong female characters are also central to Lúcio Cardoso's *Crônica da casa assassinada* (1959; *Chronicle of the Murdered House*), although their relationship is more sexually charged. As a work of high modernism, this multilayered, lengthy novel is told from the point of view of different characters and in asynchronous order. Cardoso was among the first Brazilian authors to affirm his homosexuality in his autobiographical *Diários* (1958), and the novel is among the few to address issues of gender and sexuality during this period. Most of the action takes in Minas Gerais, at the Meneses mansion on a plantation that symbolizes Brazil's traditional patriarchal, sovereign power.

Although the novel opens with the return to the Meneses estate of a middle-aged woman named Nina, we learn that she first arrived there years ago when she married the aristocratic Valdo Meneses, whom she subsequently betrayed when both his brother and the gardener Alberto fell in love with her. Following her affair with Alberto, Nina became pregnant and abandoned the family with Ana, her plain, quiet sister-in-law, who followed her to Rio de Janeiro as a way of making peace in the family. Ana eventually returned alone, bringing back a baby boy, André, whom everyone assumed to be Nina's son. At first blush, it appears that Nina has ruined the reputation of the Meneses family, but later we learn a more complex truth when Nina returns years later, her body riddled with cancer. Despite their estrangement, Valdo accepts her return to the mansion, where the young André falls in love with her and the two have what appears to be an incestuous relationship. After Nina's death, the truth is finally revealed: André is not Nina's son, but Ana's, who had hidden her own affair with the gardener. Nina actually raised her son elsewhere, far from the curse of the Meneses patriarchal mansion.

The immunological crisis is most evident in Nina's cancer, as if the house has transmitted its bad blood to feed on her. In his article "A música do sangue" (Blood music), Mário Carelli calls attention to the vampiric motifs of the novel and how the patriarchal traditions and economic and social models of plantation life drain the life out of its residents. As Carelli has observed, the home is surrounded by sinister plants. In fact, the presence of mushrooms and mold growing in the house are associated with Nina's cancer, which Cardoso describes as an illness that "ia se alastrando pela sua carne, e abrindo quenas ilhas róseas, e canais escuros, e veias que se levantavam intumescidas,

e roxas áreas de longos e caprichosos desenhos, toda uma geogra-
fia enfim da destruição lenta e sem remédio" (444; spreads throughout her flesh, opening rosy islands, dark channels and raising swol-
len veins and purple areas of whimsical design, all a geography of a
slow and hopeless demise). The mushrooms that erupt in the fields
parallel the growth of Nina's cancerous tumors, while the color of
Nina's favorite flowers, purple violets, is similar to the color of the
coagulated blood on the sheets from her bleeding tumors. All veg-
etation and flowers become leitmotifs of the decadence and decay
of the tropical mansion that represent the autoimmune disease of
a repressive society.

According to Guy Besançon, although we are led to believe that
incest is the cause of the family's demise, the actual repressed subtext of the novel is the sister-in-law Ana's desire for Nina (695). Thus,
we have a recurrence of the classic theme of lesbian desire found in
other twentieth-century gothic novels by Du Maurier and Jackson,
in which the gothic house and a vampiric presence conceal same-sex
desire.[8] In *Crônica da casa assassinada*, the parallel lives of Nina and
Ana, whose names are remarkably similar, represent the uncanny
doubles of the female gothic in a vampiric relationship. It is Nina's
courage and defiance of the gothic mansion and its patriarchal power
that allows Ana to survive as if living off of her sister-in-law's energy.
Ana's tribute to Nina, the only person strong enough to confront the
patriarchal order, is her repressed admiration and desire. Nina and
Ana both dare to challenge social conventions in having an affair with
the gardener—a man of a lower social class—but both women pay
a high price for their defiance.

While Besançon interprets the women's actions and forbidden
desire within the parameters of Freudian hysteria (694), I think it is
more useful to analyze the story in light of the theory of the gothic as
developed by Pauline Palmer. In "The Lesbian Gothic: Genre, Transformation, Transgression," Palmer notes that silence and the uncanny
tend to characterize female subjectivity and desire in the traditional
gothic text—which also characterizes the meek Ana in *Crônica da casa
assassinada*. Rosario Arias cites the example of silent rage in *Rebecca*
as well, in which the cold housekeeper, Mrs. Danvers, expresses a

lesbian desire for her imperious former mistress by projecting her hatred onto her timid new mistress, to the point that she eventually burns down the mansion and dies in the fire (51–56). The theme of silence also appears in Cardoso's novel, since the husband's homosexual brother, Timóteo, is silenced and remains closeted and confined to his bedroom. At the end of the novel, during Nina's funeral, Timóteo, dressed in women's clothes, leaves his room, thus revealing the secret of his homosexuality to the conservative rural community.[9]

Ana, who at first appears to play a minor role in the narrative, is also associated with silence and plant imagery suggestive of the disease that affects the house. When the gardener chooses Nina over her, she compares her jealousy to the spread of fungi: "O próprio calor da colcha fazia crescer em mim as imagens terríveis, obscenas: de todos os lados como fungos que brosassem da sombra" (344; The very heat of the mattress made terrible images grow within me, obscene, from all sides, like fungi that grow in the shadow). Later, when she takes Nina's blood-covered sheets to be washed, she becomes intoxicated by their scent, commenting "era como se eu estreitasse um ramalhete das mais frescas rosas, e sentisse através do seu bolo ensanguentado, não a vingança que exprimiam, mas um odor carnal e exitante de sangue e primavera" (475; it was as if I held against myself a branch of the freshest roses, and smelled in their bloody mass, not the revenge they expressed, but an arousing carnal odor of blood and spring). Her reaction to the smell is to whirl around the sheets as if dancing. While at first these actions appear to be her gloating at Nina's demise, it is actually in this moment that Ana openly expresses her passion and desire for Nina.

Although Nina's cancer can be explained in naturalistic terms, the constant comparisons of the disease with the mansion's fungi and flora lend it a symbolic value. It is Nina's distinct odor and blood on the sheets that identify her as a sexual initiator who violates marriage, social class, and age taboos. Just as her cancer—a kind of autoimmune disease — attacks the body, the parasitic plants and mushrooms attack the house as symbols of her defiance of patriarchal authority. This is the threat that the patriarchal household represses, since by defying sexual and social roles, the lesbian vampire is the ultimate taboo breaker: "Lesbian vampirism . . . is doubly abject because woman, already more abject than man, releases the blood of another woman" (Creed 61). If we apply this to Cardoso's novel, lesbian vampirism is represented symbolically in the dance that Ana performs with Nina's bloodied sheets in lieu of in actual vampiric feeding.[10] Even Nina herself is described as a giant poisonous plant: "uma imensa girassol, secreto e envenenado, alimentava-se na sombra" (302; an immense sunflower, secret and poisoned that nurtured itself in the shadow). Mário Carelli sees a parallel with vampirism in this metaphor: "A flor metafórica recebe caracterizações desmedidas; o autor tira um efeito quase oximórico do contraste entre seu aspector solar (girassol) e sua metamorfose em uma *planta carnívora ou vampiro vegetal* . . . , nocturno e secreto [grifo nosso]" (728, my emphasis; The metaphorical flower receives excessive characterizations; the author draws an almost oxymoronic contrast between its solar aspect (as a sunflower) and its metamorphosis into a *carnivorous or plant-like vampire* . . . , nocturnal and secret.) Thus, Nina takes on the characteristics of a vampire in order to defeat and kill the house that threatens her but ultimately succumbs to its bad blood. Although Nina dies, her double, Ana, survives, as if inoculated by Nina's ordeal.

Cardoso's *Crônica da casa assassinada*, published in 1959, is pioneering in its portrayal of homosexuality and same sex-desire, themes that become de rigueur in the novels of Ann Rice and films like *The Hunger* (1983).[11] During the first half of the twentieth century, vampires are generally associated with sexuality, murder, and the gothic milieu, representing collective history and the weight of the past. Vampires do allow for gender fluidity, with women challenging the patriarchal structures that oppress. Generally, they do not succeed in transforming society, but rather sacrifice themselves in the expression of their sexuality or independence. The vampiric feminine threatens to undermine national myths of patriarchal structures that feed off the energy of repressed Others, provoking the gothic mansion's autoimmune response symbolized by cancer and the plant life that feeds off its own hosts or inhabitants. The maid and the housewife in Dávila's "El huésped" work together to achieve the containment of the vampiric presence, while the attack on the gothic mansion by the women of *Crônica da casa assassinada* provokes an autoimmunological response that brings down the household. The patriarchal urge to control women's bodies recalls the violence of a slave-based society that depended on the control of bodies and discourse. As Jacob C. Brown has shown, at the heart of vampire literature is the repressed memory and violation of black women (6). In *Crônica da casa assassinada*, I would argue that the gardener Alberto is a laborer who recalls this rural slave-based past, while in "El huésped," the maid Guadalupe and her son play a similar role. Thus the tropical gothic mansion is haunted by the repressed violence against those who resist the vampiric regime of the patriarchal colonial past.

Another story of female resistance is Gabriela Rábago Palafox's "Primera comunión" (First communion) from her vampire story anthology *La voz de la sangre* (1990). The story is divided into four parts and addresses the female body and autoerotic and erotic desire explicitly. The first part portrays Sor Ángela gazing at herself in the glass of a door in her convent, admiring her own physical beauty and sensuality. The second begins with Sor Ángela taking part in a church procession as she feels menstrual blood running down her legs. She subsequently dreams that the Christ of the crucifix comes to life and forces her to drink from his wounds. In the third part, she observes two nuns at dinner intensely gazing at each other and "las sintió prometerse la noche y éxtasis" (24; she felt them promise one another a night of ecstasy). She imagines herself witnessing the promised encounter, inadvertently dropping her fork and disturbing the silence. In the fourth and final part, while sitting in her cell she hears "las voces de las hermanas que se funden, indiscretas, al otro lado del muro" (26; the voices of the sisters fusing together, indiscreetly, on the other side of the wall). Later on, in the church, she feels the wounds of the statues calling to her, inviting her to "lamer— lentamente primero, y después con avidez—las heridas que se le ofrecen" (27; lick, slowly at first, and then greedily—the open wounds they offered her). What is most relevant to Rábago Palafox's story is the ambiguous gender identity of the vampire figure and its sexual implications. Unlike Stoker's vampires, who cannot see themselves in mirrors, Sor Ángela contemplates her own reflection the glass door, which allows her to focus on her own teeth, which are "parejos y brillantes" (22; even and white), reminding us of their centrality in vampire imagery. Phrases from prayers such as "El señor está ahí. Y te llama" (The Lord is here. And he is calling to you) begin to seem ambiguous, as if she is being called not by Christ but by the vampire. Although both male and female bodies of saints call to her, the final image of her sexual impulse, "to lick" as a type of "first communion," indicates that same-sex desire is at play, and that this gesture is her first awareness of it. "Primera comunión" breaks many religious and sexual taboos and transforms bad blood into an inoculation against patriarchal control. As one of first Mexican vampire stories to refer openly to lesbian desire, it uses the baroque art of the church to personify the body, desire, and female resistance.[12]

Bíos and Citizenship: The Vampire and the Social Other

The twenty-first century witnesses, in addition to the return of the parasitic vampires of the past, the rise of a new kind of vampire who initiates a paradigm of reverse colonization and resistance. Freed from the gothic mansion, this New World vampire acts as an agent of the oppressed. The first two vampires to be discussed are also "new" in terms of origins: the vampire in Gerson Lodi-Ribeiro's *O vampiro da Nova Holanda* (1998) is an alien, while the creatures in Gabriel Trujillo Muñoz's *Espantapájaros* (1999) are bioengineered bats. Both experience persecution and violence by humans early in life but later manage to adapt to and live in cooperation with humans.

Lodi-Ribeiro's vampire saga, the first part of which won Brazil's Nova Prize for the best science fiction story in 1996, was published in Portugal in 1998 under the title *O vampiro de Nova Holanda* (The vampire of New Holland) and later appeared in the 2006 collection *Outros mundos* (Other worlds). It has subsequently been expanded and republished under the title *Aventuras do vampiro de Palmares* (2014; Adventures of a vampire from Palmares). Called *os filhos da noite* (children of the night), these New World vampires are portrayed as an ethnic group from Cuzco, who are hunted to near-extinction by the Incan crown after they overfeed on the human population. The sole survivor, known later as Dentes Compridos (Long Teeth), spends many years in isolation before crossing the continent into the territory that would eventually be known as Brazil. In this alternate world, the current geographical area that comprises Brazil is divided among three powers: the mainly Afrodescendant and indigenous Republic of Palmares, the Portuguese state known as Brazil, and the Dutch New Holland.

As Dentes Compridos travels eastward, he makes the same mistake of overfeeding, leading to his capture by the Palmarinos (i.e., former indigenous and African slaves and their descendants), who, rather than kill him, force him to act as their secret agent. Since he now works on behalf of the historically downtrodden, Dentes Compridos does not need to seek revenge on the colonizer as Dracula did. He learns to control his feeding habits so that he can coexist peacefully with humans. The Palmarino vampire learns to respect the rights of his fellow citizens, taking just enough blood to allow his "victims" to continue to live and advancing the cause of those who have few political rights under colonial rule.

An important chapter of Dentes Compridos's story involves his helping solve a series of murders in London attributed to Jack the Ripper, in the story "Assessor para assuntos fúnebres: aventuras de um vampiro palmarino em Londres" (1999; Aide in funereal affairs: The adventures of a Palmarino vampire in London). It is no coincidence that Lodi-Ribeiro's vampire, like the protagonist of Stoker's *Dracula*, appears in the capital of the British Empire, as his trajectory effectively contrasts the British colonial enterprise with that of the Portuguese. Although it is suspected that Jack the Ripper is a vampire, he turns out to be a werewolf. After being tracked down and trapped by Dentes Compridos, the werewolf explains that he is also a stranded alien, driven to commit heinous crimes out of rage and frustration at his failing strength. He wonders why the vampire appears to be getting stronger with age, while he himself is dying. Dentes Compridos, who has no mercy for the Ripper's condition, retorts, "fui tão prisioneiro das circunstâncias quanto você. Só que optei por fazer alguns amigos entre os vidas-curtas, ao invés de apreciar o nascimento e morte das culturas de cima de um pedestal inatingível." (103; I was as much a prisoner of circumstance as you. It's just that I opted to make some friends among the short-lifes, instead of observing the birth and death of cultures from a high pedestal). The vampire, by making friends with humans, imitates the Portuguese colonial experience in Brazil, in that the Portuguese did not reject or fear the racial Other, but pursued miscegenation, while the Ripper/werewolf represents the colonial pattern of the British, who maintained a more xenophobic attitude, eschewing racial mixture in favor of racial purity. A fight ensures and in the end, despite the werewolf's greater physical strength, Dentes Compridos defeats him.

Although both Stoker's and Lodi-Ribeiro's vampires fulfill the function of what Stephen D. Arata calls "reverse colonization," there are several key differences between the Old World Dracula and the New World Dentes Compridos. Stoker's Dracula, after arriving in London, begins to conquer the domain of the former colonizer by seducing aristocratic women and turning them into vampires. Arata believes that, in general, "these narratives provide an opportunity to atone for imperial sins, since reverse colonization is often represented as deserved punishment" (623). One of the main fears symbolized by Dracula is that of the mixing of blood, or miscegenation, evidenced in the bloodsucking of the vampire. Arata continues: "if 'blood' is a sign of racial identity, then Dracula effectively deracinates his victims. In turn they receive a new racial identity, one that marks them as literally Other" (630). England's fear of racial weakening and mixture is the opposite of Brazil's historical experience, and Lodi-Ribeiro's narrative does not condemn miscegenation but celebrates it. In "Assessor," the werewolf, weakened by its insistence on maintaining distance from humanity, is portrayed as outmoded because he cannot draw new vitality from it. As an ally of oppressed groups, Dentes Compridos can criticize Old World colonization as predation, because his experience is more assimilationist. The New World vampire rarely kills his victims, preferring to live among them.

Jacob C. Brown rightly criticizes the Freyrian notion of *mestiçagem* in Lodi-Ribeiro's works, noting that the saga glosses over the historical violence against Brazil's Afro-descendant women. In *O vampiro de Nova Holanda* (1998), Dentes Compridos feeds on and kills a female companion of the Bantu princess Amalamale, prompting her to hunt him down for punishment. She eventually finds and kills him, but as a supernatural being, he is able to revive. Brown identifies the vampire's excuse for committing the crime as the sexist/racist idea that black women are sexually irresistible (6), which reinforces the hidden violence of *mestiçagem*. Based on this incident from *O vampiro de Nova Holanda* (1998), along with other events from Ivan Jaf's young-adult novel *O vampiro que descobriu o Brasil* (1999; The vampire that discovered Brazil) and Nazarethe Fonseca's *Dom Pedro I, vampiro* (2015), Brown affirms that: "In all three novels, the black women have no names and no voices but their screams" (21). While I agree that *mestiçagem* tends to hide this repressed violence, I would add, to be fair, that Lodi-Ribeiro also includes strong black female characters like the Bantu princess Amalamale as well as an important black female historical figure, Chica da Silva, who appears in the story "O capitão diabo das Geraes," included in *Aventuras do vampiro de Palmares*. By emphasizing black women as both victims and victors, Lodi-Ribeiro attempts to imagine agency over victimhood in his alternate histories.

In "O capitão diabo das Geraes" (2014; Captain devil of the Geraes] (2014), Lodi-Ribeiro's most recent addition to the saga of Dentes Compridos, the vampire works against Portuguese colonialism in the eighteenth century, when the discovery of gold and diamond mines transformed the economy of the state now known as Minas Gerais. Chica da Silva (ca. 1732–1796) is a former black slave who has married the Portuguese agent entrusted with the shipments of diamonds back to Portugal. Naturally, Dentes Compridos seeks to obtain the diamonds for Palmares, and Chica secretly attempts to support the interests of the black republic against her own Portuguese husband, who in turn seeks to undermine these interests. Caught in a trap while stealing the diamonds, Dentes Compridos is hanged and appears to be dead. However, he is revived when a shaman puts his blood and that of his private guard into the vampire's mouth. Although both the vampire and Chica da Silva are initially unsuccessful, we know that their efforts will not ultimately be in vain, since in this alternate history the future República dos Palmares will gain in strength and eventually become the equal of Dutch Nova Holanda and Portuguese

Brasil as the series continues into the nineteenth century. Most importantly, Chica da Silva is portrayed as a spy for the Palmarinos, using her power and knowledge to further the anticolonial interests of the vampire and the African settlement of the Palmarinos.

The fact that Dentes Compridos is portrayed as facilitating or conferring a sense of equality and citizenship offers a perhaps facile or paternalistic inoculation that overlooks the politics of race and class that continues to plague contemporary Brazil. As mentioned earlier, Brown convincingly argues that the undying and undead vampire conveys the potency of the myth of racial democracy in Brazil. However, there is no single type of vampire in Brazil or Mexico, such that humans who attempt to resist its monstrous power are sometimes strengthened and sometimes victimized. For this reason, the immunity paradigm is a valuable tool for examining the varieties of vampires in both countries.

Gabriel Trujillo Muñoz's *Espantapájaros* (1999; Scarecrows) is also a commentary on colonial and neocolonial history, specifically the relationship between the United States and Mexico. The novel uses bat-like creatures to explore issues surrounding the border area between the two countries, including illegal immigration and drug trafficking. Similar to Lodi-Ribeiro's vampire series, *Espantapájaros* has a utopian impulse, imagining an alternate future that includes an intelligent, biologically engineered species of bat that can also be construed as vampiric.

The novel is a cross between science fiction and espionage thriller, since it involves a secret biotechnology project established after World War II for the creation of vampire-like creatures to be used to gather military intelligence. The goal was to produce "un ser volador indetectable, capaz de tomar decisiones propias, con una visión extrema, que pudiera infiltrarse en territorio enemigo e informar sobre movimientos de tropas, y en caso necesario, proceder a destruirlas" (50; an undetectable flyer, capable of making its own decision, with extremely sharp vision, able to infiltrate enemy territory and give information about troop movement, and kill them, if necessary). Known as Gracos (an acronym for Genetic Research for Animal Conscience and Operativity), these creatures end up developing their own language, unbeknownst to their creators. By 1995, they manage to escape from the laboratory and begin feeding on cattle in Arizona. The US Army, while publicly dismissing reports by farmers as attacks by mythical *chupacabras* orders Commander Duloth, who had an important role in the creation of the Gracos, to engage in a secret mission to contain the threat they represent. He reluctantly accepts the assignment, as he will be forced to enact the immunological paradigm of containment and deportation, recalling the internment camps and other features of the immunological response.

However, the vampires prove to be a formidable enemy, since in one battle only three Gracos perish while 152 human soldiers lose their lives. We learn that the Gracos can fly for weeks, that their bite emits a hypnotic and hallucinogenic substance, and that they are incredibly fertile. As oviparous hermaphrodites, they can reproduce three to four times a year. At this point the Gracos begin to resemble Stoker's vampire as a challenge to the hegemonic culture of humans, as described by Arata: "The racial threat embodied by the Count is thus intensified: not only is he more vigorous, more fecund, more 'primitive' than his Western antagonists, he is also becoming more 'advanced.' . . . Dracula's swift developments will soon make him invincible" (639–40). After the initial stunning defeat, the army adopts a goal of annihilating the Gracos, and the full-scale immunological paradigm comes into view. When considered in the context of the border between Mexico and the United States, we can see the implications of the story in its depiction of an immunological response against an Other that is willing and able to mount a counter-immunological defense. *Espantapájaros* also draws clear parallels with the treatment of Mexican immigrants and workers in the United States: after a period of cooperation through the *bracero* program, which legalized immigration during World War II, a more hostile attitude toward illegal immigration became prevalent as Americans' fear of job loss and swelling social welfare costs began to strain border relations. In Trujillo's novel, the Gracos and the Mexican immigrants become allies, as shown when a Graco saves a Mexican worker trying cross the desert into the United States.

Months of bloodshed ensue, but when one of the creatures leaves its dead offspring in the arms of a scarecrow—hence the name of

the novel, *Espantapájaros*—this is interpreted as an offer of reconciliation between the two species—a gesture that employs an image of
both birth and death. This leads to what Esposito calls a glimpse of
"another possibility, tied to a different mode of understanding in the
relation between the phantasm of death and the power of life" (*Bíos*
169). Duloth gradually begins to understand and respect the Gracos,
attitudes for which he is later charged with treason. As in the case
with Lodi-Ribeiro's alien vampire, by the end of the military operation against the Gracos, only one young Graco escapes alive. He is
able to grow to maturity with the help of a Mexican woman who
had suffered a series of miscarriages. She and her husband name the
creature Ángel and raise him on milk, oil, and fruit, so that he will
not learn to feed on blood and threaten humans. Hence, he also controls his appetites, like Dentes Compridos, in order to begin a process of reconciliation. The use of the Christian concept of "angel" to
describe a creature forged by biotechnology seems to capture some
of Esposito's ponderings, as he states that "One could say that biotechnology is a non-Christian form of incarnation" (*Bíos* 168). In the
case of the biopolitics of the Gracos, the question is whether their
rights as a conscious life form will be recognized. And indeed, when
the single remaining Graco manages to replicate itself many times
through self-fertilization while receiving protection from an underground ecological movement, its descendants eventually achieve
recognition from the United Nations as a third intelligent species
(after humans and dolphins), with protected status and the right to
live among humans.

Lodi-Ribeiro and Trujillo Muñoz both use vampires in utopian,
almost didactic tales of reverse colonization. In the Brazilian alternate history, the presence of the vampire levels the playing field so
that the descendants of former slaves and indigenous peoples are
able to create their own countries within Brazilian territory and resist
exploitative European patterns of colonization. The Mexican narrative abounds in critiques of American imperialism, including commentaries on the injustices of the Mexican-American War, the unfair
treatment of Native Americans, and the American brutality in the
pursuit of Pancho Villa, and even more contemporary issues such

as the Gulf War, the War on Drugs, and immigration. The vampire becomes a figure of racial and cultural mixture, a truly alien, intelligent species capable of threatening humans (Americans) but choosing to share the planet with them. For Ramón López Castro, the premise of *Espantapájaros* is original: "Muy difícil encontrar una novela que proponga una ruptura conceptual tan amplia como la de *Espantapájaros*" (174; It would be very difficult to find a novel that proposes a conceptual rupture as radical as that of *Espantapájaros*), since it uses biotechnology in order to challenge in a radical way our concepts of humanity, gender, reproduction, intelligence, species, and race.

The novella, which is Trujillo Muñoz's best-known work of science fiction (Lockhart 201), won recognition in Spain under the title *Gracos* in 1998, then won the Mexican Premio Nacional de Narrativa Colima (Colima Prize for National Narrative) the following year. As a novel that attempts to allegorize border relations, racial fear, and resentment, it successfully incorporates the vampire figure into science fiction. Although it emphasizes issues of race and prejudice, the fact that the Gracos are oviparous and hermaphroditic suggests new paths in gender relations and gender identity as well. This is an interesting aspect of the novel that is not developed further, suggesting the gender fluidity of the vampire figure as an alternate model of the body, suggestive of Esposito's notions of birth and becoming (*Bíos* 176), while exploring the tensions of the border and of interspecies relations.

While Mexico has produced few vampire novels in recent years, they have proliferated in Brazil. This increase can be largely associated with the 1999 publication of *Os sete* (The seven) by André Vianco, an author who self-published the novel before it was picked up by Novo Século in 2001.[13] Briefly, *Os sete* manages to Brazilianize the vampire for young adults while touching on crime and other ills of urban life, both predating and then riding on the success of *Buffy the Vampire Slayer* and the *Twilight* saga of the early 2000s.[14] As a novel of almost four hundred pages, *Os sete* has a complex plot that centers on Tiago, who becomes a human/vampire hybrid and manages to defeat seven Portuguese vampires, each endowed with supernatural powers.[15] The vampires make patent the inability of Brazil's traditional

institutions—the church, the university, the army, and the police—
to deal with issues of crime and injustice effectively. As long as he
refrains from eating meat, Tiago will not become a full vampire and
will defend humanity. He succeeds in his plan to expel the vampires
by luring them to the ocean and involving them in a fatal explosion.
The sense of insularity or protection is key here, because Tiago must
protect Brazil, closing it off from the colonial vampires.

Two texts that attempt to extend the immunological paradigm to
the Other include José Luis Zárate's *La ruta del hielo y la sal* (1998) Giulia
Moon's *Kaori: Perfume de vampira* (2009). Both novels feature characters who, despite being vampires or being bitten themselves, reach
out beyond their own interests. These alliances among humans and
the undead open the body and the body politic to new configurations.
Donna Haraway has written that rather than protection, "Immunity can also be conceived of in terms of shared specificities; of the
semi-permeable self able to engage with others (human and nonhuman; inner and outer) . . . [that] must be brought into other emerging Western and multi-cultural discourses on health, sickness, individuality, humanity and death" ("Biopolitics" 299). Both Zárate and
Moon allude to this "semi-permeable" sense of self through human/
vampire interactions.

Paying tribute to Bram Stoker's *Dracula*, José Luis Zárate's *La ruta
del hielo y la sal* (1998; The route of ice and salt) bases its action on a
chapter of Stoker's novel, "Log of the Demeter," in which a Russian
ship transports Dracula's body from the port of Varna to Whitby,
England.[16] In Zárate's *La ruta del hielo y la sal*, the Russian captain is
a homosexual plagued by sexual anxiety in late nineteenth century.
In his log, he does not refer to "Dracula" or "vampires," but he does
write about an evil presence aboard the ship that affects his mind
and body and those of his crew. *La ruta del hielo y la sal* is mainly an
interior monologue by the captain on the sexual desire he feels for
the bodies of his crewmen and his struggle to repress these feelings
and his memories of his young lover. Like the ship's log in *Dracula*,
Zárate's log describes the loading of the ship, the nine crewmembers, and the strange cargo of large silver boxes seemingly filled
with dirt and labeled as material for scientific experiments. Once the

boxes are aboard, the captain is plagued by dreams of bats, alarmed by the unexplained opening and closing of the crates, and puzzled by the appearance of a bloody mark on a crewmember's neck. He also suffers from delusions of rats touching his body sexually and of the ship becoming his sexual partner. In the second part of the novel, chaos descends upon the ship and the captain records that some crewmen kill themselves in terror while others succumb to attacks by the vampire.

At the same time, the captain and the vampire are portrayed as doubles. For example, at the outset of the trip, when the captain is gathering his crew, he states, "siempre es necesaria la sangre nueva" (16; new blood is always necessary), a metaphorical first reference to blood. When his finger is pricked by a splinter from one of the boxes he is examining, he recalls the time his lover Mikhail kissed and then bit him (44). There is also a reference to the coffin-like dimensions of the captain's quarters, and in one instance he compares his homosexual desire to a vampire's hunger for blood. The captain also believes that he was responsible for the death of his lover.

By the third part of the story, the only remaining crewmembers are now themselves vampires who long for the captain's blood yet dare not approach him because of his rank. He learns from one of them that he did not kill his lover, which lessens his sense of guilt. The captain then resolves to save himself spiritually and to do everything possible to prevent Dracula's arrival in England. He destroys all of his navigational aids, but knows that he also needs to eliminate the remaining sailors. Since he knows that vampire bodies will dissolve like salt in water and in the light of the sun, he jumps into sea, tied to a rope and with a knife in his mouth. Looking back at his sailors, he regrets having brought them on this journey and realizes that he loves them and forgives them. He opens up a vein, filling the water with blood, upon which, unable to help themselves, the sailors jump in to feed and slowly dissolve. The captain offers his body and blood, Christ-like, while the sea performs a baptismal purification that allows him a sense of peace and closure, knowing that he has tried to save humanity. This salvific gesture is key because it conveys a wider message and creates a bridge between the human captain and his vampire sailors. The captain manages to climb back aboard the ship, where he ties his hands to the tiller with a rosary, knowing that Dracula will not dare to come near it. He keeps the log in a jar in his belt, just as in Stoker's novel. The captain's final interior monologue conveys his sense of inner peace for having reached out to his crew.

The captain demonstrates a desire to reconnect with both vampires and humanity by making himself vulnerable and opening up his body to others. Zárate's hero has much in common with Juan Gris, the protagonist of Guillermo del Toro's film 1993 *Cronos*, in which the hero makes a similar gesture of sacrifice. As Persephone Braham notes: "*Cronos* not only refuses to observe the dichotomies that justify definitions such as monstrosity and humanity, but insists on the legitimacy—even the sanctity of the vampire hero" (*From Amazons* 174–75). The captain of Zárate's novel overcomes similar dichotomies and shows vulnerability and heroism in choosing to sacrifice himself, becoming, like *Cronos*'s Juan Gris, a vampire hero who serves humanity.

Among the plethora of vampire works published in Brazil, Moon's *Kaori: Perfume de vampira* (2009; Kaori: Scent of the vampire) is exceptional in touching on issues of social class, history, immigration, and alternate communities, and in assuming a more nuanced approach toward urban issues and crime. This first novel of the *Kaori* trilogy consists of two narrative threads that slowly weave together as we follow characters in twenty-first-century Brazil and the Edo period of Japan (1603–1867), complete with historical footnotes.[17] The first thread takes place in São Paulo in 2008 and follows Samuel Jouza, who works for an institute that secretly tracks and studies the activities of vampires and other beings. The second begins in seventeenth-century Japan and follows the life of a young female vampire named Kaori, who is fleeing from Missora, a female vampire who killed her father.

The historical Japanese narrative of the second thread, which includes fights, sadism, female submission, and titillating sexuality,[18] contrasts with the social portrait of contemporary São Paulo, Brazil's largest urban center. We learn that there are two groups of vampires in São Paulo, one Nisei and the other nonimmigrant Brazilian. The former are eager to live in peace with the human population, even accepting a diet of artificial blood to end the killing, but the latter, secretly led by the evil Japanese vampire Missora, are determined to continue their predation in Brazil.

Interestingly, a new form of predation is carried out by shape-shifting werewolves called *famélicos* (the hungry ones), who, manipulated by the Brazilian vampires, have begun to kill off those who observe and monitor vampire behavior for the institute. Samuel himself is injured by the *famélicos* while unsuccessfully trying to save a recently orphaned homeless boy named Davi. When Samuel joins forces with an institute biologist named Beatriz, he finds out that the *famélicos* are actually shape shifters, capable of taking human form, allowing them to blend into the urban landscape and the city's lower classes when they are not feeding. Beatriz tries to save one of the aging *famélicos* who speaks to her, explaining how he resisted the vampires' conscription of his pack. Later, Samuel is kidnapped by the Nisei vampires, who release him in a deal with the institute but not until after he has been bitten by Kaori. However, Samuel eventually helps Kaori and the Nisei vampires defeat their Brazilian foes led by Missora, destroying at the same time the syndicate that supported them.

Moon constructs a multi-tiered society of vampires and those who live in their shadow that is not typical of Brazil's popular vampire literature. In many ways, the society reflects Brazil's actual class structure: the privileged vampires function as the elite class, the institution that researches them is reminiscent of society's middle sectors, while the wolf-like *famélicos* and the street children are at the bottom of the hierarchy. The *famélicos* and the orphaned boy Davi recall João Biehl's "zones of abandonment" in neoliberal Brazil, where life is precarious (122). Samuel strives to understand both worlds, refusing to divide society into *zoê* and *bíos*, *famélicos* and vampires, or humans and monsters. Samuel and Beatriz fortify their commitment to social engagement within the wider urban space of São Paulo, acting in ways that transcend personal interests while recognizing the value of life beyond neoliberal systems of biopower that benefit the state or the market (Campbell 77–78), thus laying the foundations of an affirmative biopolitics.

Conclusion

In the course of this chapter we have seen that vampires assume many roles in Mexican and Brazilian speculative fiction. Foreign vampires invade urban homes to feed on the living, provoking a defensive immunological response. Other vampires inhabit gothic mansions, infecting with "bad blood" the lives of women who dare to challenge the patriarchal order. Vampires unexpectedly emerge in both countries as colonial avengers, championing the victims of colonial and neocolonial regimes, providing protection from necropolitical policies. Finally, we find human characters who open themselves to an alternate view of vampires in an effort to re-establish communities through mutual acceptance and a common *munus* or commitment.

The vulnerability of bodies is apparent in vampire fiction and, ironically, contact with vampires often leads characters to rethink their personal relationships and social bonds. Judith Butler reminds us that we form "socially constituted bodies, attached to others, at risk of losing those attachments, exposed to others, at risk of violence by virtue of that exposure" (*Precarious* 20). In this sense, the human struggle with vampires is a reminder both to resist seduction and violence and to reach out to each other. By exposing and expelling corruption and the secrets of the past, the vampiristic figure of the undead calls for a reanimation of the wider social struggle to transform *zoê* into the *bíos* of a functioning society.

Afterword

This study endeavors to showcase the originality and social commitment that Mexico and Brazil have brought to speculative fiction. One of my main goals, therefore, is to honor the writers past and present who have contributed to this tradition, paying tribute to the richness of Mexican and Brazilian literary history.

Taking a historical perspective and including a wide variety of texts, the study traces these two cultures' reaction to modernity through embodiment and resistance. Cyborg bodies, queer bodies, and zombified and vampiric bodies offer resistance to the positivist ethos and provide an alternate perspective on rapid growth and modernity in the twentieth and twenty-first centuries and the political and economic challenges they have brought.

The body connects us to the earth, to each other, and to society at large. Latin America's baroque ethos testifies to the resourcefulness and resilience of cultures that have long been part of capitalism but have created an alternate modernity that is perhaps most clearly exemplified in speculative fiction. While biopolitics also addresses wider issues of labor, sexuality, disease, and institutional control, the baroque ethos may be seen as the foundation of cultural immunity to necropolitical and other institutional forces. Baroque bodies may be supplemented, adorned, strengthened, or inoculated, recombining male and female, human and machine, and the living and dead as a way of questioning the binaries of our increasingly homogeneous globalized modernity.

While the altered bodies (regendered, re-engineered, and revived) that form the object of this study have historical and even mythical antecedents, each proves useful in understanding and reconstituting the history of the speculative fiction genre in Latin America. This has allowed me to reformulate the usual definition of science fiction beyond stories based on technological innovation to include bodily estrangement as well. Given the region's neocolonial history, the body (or body politic) has served as a key site for futuristic experimentation and speculation. The worldwide popularity of Netflix productions such as Pedro Aguilera's Brazilian dystopian thriller 3% confirms a growing interest in speculative fiction from Latin America.

NOTES

1. See Achille Mbembe's "Necropolitics," whose title modifies Foucault's concept of "biopolitics" to illustrate how governments use "legitimate" power not to control crime but to target specific members of the population for elimination. For examples of violence used for political control in Brazil, see Sebastián Schlofsky's 2016 article, "Policing in Two Cities: From Necropolitical Governance to Imagined Communities." For Mexico, I refer to a 2016 lecture by Ignacio Sánchez Prado, "War and the Neoliberal Condition: Death and Vulnerability in Contemporary Mexico." Sánchez Prado discussed how certain populations have been marked as "disposable" as a result of drug wars, violence in Central America, and neoliberal economic policies, as in the case of the unsolved murders of female factory workers in the *maquiladoras* (large factories on the US Mexico border). He describes the bodies of these victims as the new "product" of international trade and power relations.

2. The cultural, economic, and literary parallels between Mexico and Brazil come into sharp focus when compared to the trajectory of Argentina, a country long recognized for its contributions to the genre of speculative fiction. With its strong sense of national identity, Argentina has been shaped by distinct racial and political history, which is expressed in its cultural concerns and concepts of the body politic. Although the slave trade and the institution of slavery were long a part of the country's history, its 1813 Law of the Free Womb and the relatively early abolition of slavery in 1853 (compared to Brazil's date of 1888) minimized the impact of slavery in the collective cultural and political

imagination (E. Edwards). At the same time, the campaigns that devastated the indigenous populations in the 1870s effectively cleared Argentina's vast hinterland for European settlers, who arrived in great numbers following the 1876 immigration act, which was designed to populate the country's rich agricultural lands and cosmopolitan capital (Meter). For these reasons, the country's twentieth-century political life was shaped in a way that generally did not follow the large state projects of *mestizaje/mestiçagem* and national consolidation that took place in Mexico and Brazil. The struggles among the middle class, workers, and landowners at the beginning of the twentieth century, the divisiveness of Peronism at mid-century, and the trauma of military dictatorship have arguably lent a distinct conception of the body and the body politic to Argentine speculative literature as a battle fought more among equals.

3. The body in general terms has innumerable cultural correlates, including early political doctrines of the sovereign body, the sacred body of the Catholic Church, and ideas of the corporate state in Latin America, in addition to its biopolitical connotations. Syncretic belief systems give rise to religious narratives that invoke the sacred bodies of various indigenous goddesses or mother figures associated with national origins, from the Aztec mother Coalicue, Doña Marina (Malinche), or the Virgin of Guadalupe in Mexico, to Jaci for the Guarani, Iemanjá in Afro-Brazilian culture, and Our Lady of Aparecida, the patron saint of Brazil. In Octavio Paz's Malinche, in *El laberinto de la soledad* (1950), and José de Alencar's eponymous fictional Amerindian mother, *Iracema* (1865), we have indigenous mothers whose conquest "mythifies" the origins of the two mixed-race cultures and *mestizaje.*

4. Mexico and Brazil are the two most industrialized countries in Latin America, along with Argentina, according to Peter Evans and Gary Gereffi, in their 1982 essay "Foreign Investment and Dependent Development" (113).

5. In *Society Must be Defended: Lectures at the Collège de France 1975–1976*, Foucault outlines his ideas regarding the shift in sovereign power from the threat of the sword to the modern management of populations through biopolitics (239–63).

6. Here I take my cues from Alfred Stepan's *The State and Society* (1978), where he uses the term "organic-statist" to distinguish contemporary Latin American political organization from corporatism in Europe. In his view "there exists a recognizable strand of political thought . . . that runs from Aristotle, through Roman Law, absolutist and modern Catholic social thought . . . [that] represents powerful philosophical and structural tendencies found throughout Western Europe, and especially in the Iberian countries and their former colonies, where organic statism was never as fully challenged by alternative political models as in the rest of the European cultural areas" (4).

7. In both Mexico and Brazil, after independence in the nineteenth century, as liberals and conservatives fought among themselves and entered into wars with other countries (Mexico fought against the US and France; Brazil against

Paraguay), liberal ideologies were diluted as military officers and positivists came to power. In Mexico, the late nineteenth century is associated with the protracted rule of Porfirio Díaz (1876–1910), the "Porfiriato," and in Brazil, with the military governments that established the First Republic, with positivists and the oligarchy dominating its vision (1889-1929). One major difference is that, following the Mexican Revolution (1910–1920), the military did not assume an active political role in the twentieth century as it did in Brazil.

8. Whereas the US and Brazil both engaged heavily in the trans-African slave trade, Mexico imported relatively few African slaves, instead using indigenous labor for mining and agriculture under conditions that resembled slave-based labor but did not involve large-scale buying and selling of workers. As the colony known as "New Spain," Mexico adapted the Spanish caste system to its own situation, establishing a long and complex hierarchy from light to dark skin. See Gerardo Gutiérrez, "Identity Erasure and Demographic Impacts of the Spanish Caste System on the Indigenous Populations of Mexico" (119–45).

9. See Roberto DaMatta's 1995 essay "For an Anthropology of the Brazilian Tradition," for the contrast between the dualistic exclusionary concept of race in the United States and the triadic or intermediary concept in Brazil, which may also be similar to that of Mexico. Whereas the American ideal was "separate but equal," Brazil aspired to an ideal of "different and complementary" (272-74), an idea that is similar to Mexico's narrative of an idealized racially blended society (*mestizaje*).

10. Ben Penglase has used de Certeau in describing how residents in Rio's favelas deploy "social tactics" that consist of avoiding conflict and violence by "carving out spaces of autonomy" in order to live with constant insecurity (6-7). It is the practical application or embodied form of knowing how to "dodge, evade, or turn to one's advantage the obstacles that life [has] placed in one's path" (7).

11. Although both Mexico and Brazil were colonized in the sixteenth century, by Spain and Portugal respectively, they developed distinct national narratives. While Mexico, then known as New Spain, tells the story of military conquest and the subjugation of the Meso-American civilizations, Brazil has a more complex narrative that includes wars and alliances among diverse indigenous groups, conflicts with the Dutch and the French, and the adventures of settlers sent to colonize the interior. In Mexico during this century, the large indigenous populations were forced to labor in mines and agriculture, while the Portuguese, finding fewer indigenous workers, opted to import African slaves to work on their sugar plantations.

12. Brazil's preoccupation with its own unrealized potential or postponed utopia is perhaps most iconically represented by the founding of the capital city Brasília in 1960. Brazilian historian José Murilo de Carvalho notes that the seventeenth-century visionary Jesuit Antônio Vieira called Brazil the mystical "Fifth Empire" (23–24). Brian L. Price speculates about Mexico's ambivalent attitude toward

the past in his study *The Cult of Defeat in Mexico's Historical Fiction: Failure, Trauma, and Loss* (New York: Palgrave Macmillan, 2012). In Mexico, several events—including the Conquest, protracted struggles for independence, the Mexican-American War of 1848, the French Intervention 1862-1867, and the death of liberalism represented by Benito Juárez's demise in 1875—all contributed to what Price describes as Mexico's pact with the "transcendental value of martyrdom" (9) and its obsession with the past as seen in various literary works and debates among the Mexican intelligentsia.

13.   In addition to Vasconcelos's and Freyre's contributions to the study of national culture and psychology, many more studies could be cited. My reference to "solitude" is an obvious reference to Octavio Paz's *El laberinto de la soledad* (1950), the best known study of the Mexican character and its foundational myth of the Conquest in "Los hijos de la Malinche." Yet Paz's work belongs to a long tradition that includes works such as *Forjando patria* (1916; Forging a fatherland), in which anthropologist Miguel Gamio refutes the tradition of positivism and assimilationist views of his time as he champions indigenous cultures, and *El perfil del hombre y de la cultura en México* (1934; *Profile of Man and Culture in Mexico*), in which Samuel Ramos examines the spiritual and educational challenges of post-revolutionary Mexico. A later work on this topic is Roger Bartra's *La jaula de la melancolía* (1987; *The Cage of Melancholy*), which resists the typology of earlier studies by using the axoltl to explain Mexico's struggle with nationalism and the forces of modernity. Similar studies of Brazilian character also abound. The idea of "cordiality" was proposed by Sérgio Buarque de Holanda in *Raízes do Brasil* (1936; *The Roots of Brazil*), referring to the patriarchal concept of noblesse oblige that also masks violence and oppression perpetrated by slave owners. Other studies include Paulo Prado's 1928 *Retrato do Brasil* (Portrait of Brazil), which argues that indigenous, African, and European Brazilians all share a sense of melancholy or orphanhood; anthropologist Darcy Ribeiro's *Os brasileiros* (1995; The Brazilians), which claims that Brazilians are the "new" mixed race people of the Americas; and Roberto DaMatta's *Carnavais, malandros e heróis* (1979; *Carnivals, Rogues and Heroes*), which outlines the "triangular" nature of the Brazilian hero who is part *malandro* (or hustler), part *santo* (or saint-like) and part *caxias* (or stickler for order) as the paradigm for Brazilian culture. Oswald de Andrade's "Manifesto antropófago" (1928; Cannibalist manifesto) offers a more critical form of national identity by proposing that Brazil should "consume" the colonizer in order to form a distinct culture. Although my summaries here are cursory, some familiarity with them helps explain the cultural traditions of Mexico and Brazil that form the basis of discourses of "exceptionalism," national identity, and artistic production.

14.   Paulo Moreira and Fred Ellison have both written books exploring intellectual exchanges between Mexico and Brazil, but as mentioned, these are rare comparative cultural and literary studies. See Paulo Moreira's *Literary and Cultural*

Relations between Brazil and Mexico: Deep Undercurrents (New York: Palgrave Macmillan, 2013) and Fred Ellison's *Alfonso Reyes e o Brasil: Um mexicano entre cariocas*, translated by Cristine Ajuz (Rio de Janeiro: Topbooks, 2002). A more thematic study, Judith Payne and Earl Fitz's 1993 *Ambiguity and Gender in the New Novel of Brazil and Spanish America* compares mainstream novels of the two regions, concluding that Brazilian novels are more "gender fluid" than those of Spanish America. This conclusion is contradicted by some of the works I cite in my chapter on gender and sexuality.

15. The speculative fiction of Mexico and Brazil is often overshadowed by the realist canon, which dominates their respective literary histories. In tracing this tradition in Brazil, Flora Süssekind's 1984 study, *Tal Brasil, qual romance* (So goes Brazil, so goes the novel) outlines how late-nineteenth-century naturalism, regionalist novels of the 1930s, and journalistic novels written during 1970s document national "growing pains" and the emphasis on literary realism and national identity. In Mexico, the case is a bit different, since the "novela de la revolución" (novel of the Revolution) and testimonial novels tended to set the agenda during the decades that followed. Mexico's 1992 quincentenary produced a debate on and numerous examples of the "new historical novel." See Victoria E. Campos's 2001 chapter "Toward a New History: Twentieth-Century Debates in Mexico on Narrating the National Past."

16. See Teresa Caldeira and James Holston, "Democracy and Violence in Brazil" (1999) for the idea of failed democracy and violence against low-income populations.

17. In his essay "The Experimental Novel," Zola cites the experimental methods of physiologist Claude Bernard as the basis for his literary model of naturalism, stating, "Now, science enters into the domain of us novelists, who are today the analyzers of man, in his individual and social relations. . . . In one word, we should operate on characters, the passion, on the human and social data, in the same way that the chemist and the physicist operates on living beings" (648).

18. Calogero notes that Mexican positivism was based on Comte's earlier writings, which were more "accommodating" to the Catholic Church, while the movement in Brazil was influenced by later writings that emphasized social reform and the adoption of a religion of humanity (36).

19. For the concepts of scientific racism and doctoring of the body in Brazil, see Dain Borges's 1993 article "'Puffy, Ugly, Slothful and Inert': Degeneration in Brazilian Social Thought, 1880–1940," Lilia Moritz Schwarcz's *O espectáculo das raças* (1993; The spectacle of races) and Nancy Leys Stepan's *"The Hour of Eugenics"* (1991), which focuses on eugenics in Argentina, Mexico and Brazil. Laura Luz Suárez y López Guazo's study *Eugenesia y racismo en México* (2005; Eugenics and racism in Mexico) also examines the use and misuse of the "well born" science in Mexico.

20. In Mexico, muralists portrayed Mexico's past and present in grand panoramas, celebrating indigenous culture and popular art in events such as La Noche Mex-

icana (The Mexican night) and the Exhibition of Popular Arts in 1921, marking the centenary of independence (Rick López 3–40). The poets and artists of São Paulo's Semana de Arte Moderna (Week of modern art) in 1922 celebrated Brazil's popular roots, and Cândido Portinari's large-scale paintings of Brazil's popular classes, which can be contrasted with the modernist and futuristic esthetic of Oscar Niemeyer's architecture, illustrated the contradictions of the tradition and the modern. Musical nationalism was promoted by composers such as Carlos Chávez in Mexico and Heitor Villa Lobos in Brazil, both of whom reworked popular folksongs into national and classical forms. While Mexico's capital renamed its streets in honor of its history and the Revolution and popularized the Day of the Dead and other celebrations, Brazilian cities opened their streets to carnival parades. In sports, in 1932 President Getúlio Vargas legalized the formerly forbidden Afro-descendent practice of *capoeira*, a combination of dance and martial art, calling it the first genuine Brazilian sport, and in 1938, his government recognized the value of promoting Brazil's professional and national soccer teams and the Afro-descendant stars who helped them achieve spectacular victories (Kittleson 34–37). Concurrently, Mexican wrestling emerged in its unique form of masked *luchadores* (wrestlers) in 1933 when Samuel Lutteroth founded the Empresa Mexicana de Lucha Libre (Mexican wrestling corporation), a league whose stars were later popularized in television and film. In this way, the arts and sports featured *mestizo/mestiço* bodies as the symbol of consolidated modern national identity.

21. Fred Ellison opens his study *Alfonso Reyes e o Brasil* (2002; Alfonso Reyes and Brazil) by recounting José de Vasconcelos's 1922 visit and its positive repercussions for his vision of relations between the two countries (24–27). As members of the Ateneo de la Juventud (Atheneum of Youth) (1909–1914), a group that promoted a series of cultural debates and philosophical discussions about education and Mexican cultural identity, Vasconcelos and Freyre shared a deep interest in pan-Latin American culture, despite their later political differences (27–28).

22. For a detailed description of the four *ethe* in English, see Stefan Gandler, *Critical Marxism in Mexico*, 295–305.

23. Unless otherwise indicated by a page number or an entry in the Bibliography, the translations of texts and titles from Spanish and Portuguese into English are my own.

24. Echeverría notes that when a subaltern says "yes," there is a resistance because of the power difference: "El subordinado está compelido a la aquiescencia al dominador: no tiene acceso a la significación 'no.'" ("Ethos" 35–36; The subordinate is compelled to acquiesce to the more powerful: he does not have access to the meaning of "no").

25. Gandler has summed up Echeverría's thoughts as follows: "The folklorization of Latin America's 'baroque' or 'magical realists' seems to Echeverría to be, in the final instance, nothing more than a part of the strategy of the dominant

ethos to rid itself of the baroque ethos by banishing it to the 'non-world of pre-modernity'" (307), or in Echeverría's words: "Substantivar la singulari-dad de los latinoamericanos, folclorizándolos alegremente como 'barrocos', 'realistas mágicos' etcétera, es invitarlos a asumir, y además con cierto dudoso orgullo, los mismo viejos calificativos que el discurso proveniente de las otras modalidades del ethos moderno ha empleado desde siempre para relegar al ethos barroco al no-mundo de la pre-modernidad y para cubrir así el trabajo de integración, deformación y refuncionalización de sus peculiaridades con el que ellas se han impuesto sobre él." ("Ethos" 28; To concretize the singu-larity of Latin Americans, folklorizing them happily as "baroque," "magical realists," etc., is to have the concept that—even with a certain pride—they belong to same old descriptions that the discourse from the other modalities of the modern ethos has always used to relegate the baroque ethos to the non-world of pre-modernity and to cover up, in this way, the work of inte-gration, deformation, and refunctionalization of the peculiarities that have been imposed on it.)

26. Gandler makes this observation about attitudes toward revolutionary action in Echeverría's writings (233n130).

27. Perhaps the fundamental difference between Andrade and Echeverría is gener-ational. Whereas Oswald de Andrade's generation still believed in revolutionary ideals and the demise of capitalism, as he portrayed in the utopia of the matri-archal Pindorama (the Tupi name for Brazil) in *Do Pau Brasil à Antropofagia e às Utopias* ("À marcha das utopias," 208–9). Echeverría was affected by events of the 1990s, namely the advent of NAFTA and the Zapatista movement, hence his project of reimagining popular resistance to capitalism.

28. As Elaine Showalter notes in her 1990 study *Sexual Anarchy: Gender and Culture at the Fin de Siècle*, homosexuality was outlawed in 1885 under the Labouchère Amendment in England, whence the imprisonment of Oscar Wilde (14). Homo-sexuals were often forced to lead double lives, leaving them open to blackmail (112).

29. Macías-González cites the 1901 scandal of the "41," in which homosexual men attending an all-male gala stepped over the line between public and private codes of behavior. He notes that while the event did not cause a "legislative backlash," the accused were forced to sweep streets and spend time in jail (133–34). For an in-depth discussion of press coverage, see Sifuentes-Jáuregui's 2002 study *Transvestism, Masculinity and Latin American Literature*, 15–51. Robert McKee Irwin's 2003 *Mexican Masculinities* examines the event's cultural aspects, including analysis of a *corrido* (popular song) mocking the event, with images by cartoonist José Guadlupe Posada, and of a novel, *Los cuarenta y uno* (The forty-one) by Eduardo A. Castrejón (pseud.), 78-91.

30. Esposito explains that the immune system generally operates on the principle of "auto-tolerance," though it sometimes ends up attacking the very tissues it is designed to defend (164-65).

31. I am referring to Timothy Campbell's interpretation of Esposito's concept of the "impersonal," by which he means the "sacred" or the shared elements that go beyond personal interest: "To think the impersonal . . . breaks with what is considered ours and what is considered mine. . . . Clearly I am pushing Esposito's thought here in directions he does not explicitly acknowledge; his object is less a distancing from the thanatopolitical than it is a laying of the foundations of an affirmative biopolitics" (78).

32. In English, the following materials are available: The science fiction traditions of Latin America have been mapped out in the country-by-country bibliographies elaborated by Molina-Gavilán et al. in the 2007 "Chronology of Latin American Science Fiction." For a detailed history of Brazilian science fiction, see Ginway and Causo, "Discovering and Rediscovering Brazilian Science Fiction: An Overview," and for Mexican science fiction, see Ross Larson's *Fantasy and Imagination in the Mexican Narrative* (1977) and Darrell B. Lockhart's *Latin American Science Fiction: An A-to-Z Guide* (2004). The introduction to Andrea L. Bell and Yolanda Molina Gavilán's *Cosmos Latinos: An Anthology of Science Fiction from Latin America and Spain* (2003) offers essential bio-bibliographical, historical, and thematic surveys of science fiction and fantasy in Latin America in general. For an overview of Mexico and Brazil, see the entries in the online version of Clute, Langford, and Nicholls's *Encyclopedia of Science Fiction*. The entry for Mexico is by Miguel Ángel Fernández Delgado, for Brazil by Braulio Tavares and Roberto de Sousa Causo.

33. Regarding Brazilian twentieth-century utopian literature, in addition to Haywood Ferreira's analysis of "Brasil no ano 2.000" by Godofredo Emerson Barnsley, see Ginway, "*O presidente negro*" and "Eugenia, a mulher e a política na literatura utópica brasileira (1909–1929)." These works illustrate the conservative (and openly racist) nature of Brazilian utopian literature, which contrasts with the more original and progressive elements of Urzaiz's 1919 *Eugenia*.

34. For readers of Spanish and Portuguese, see Gabriel Trujillo Muñoz's *Biografías del futuro* (1999; Biographies of the future), Miguel Ángel Fernández Delgado's introduction to *Visiones periféricas* (Peripheral visions), Causo's *Ficção científica, fantasia e horror no Brasil 1875-1950* (2003; Science fiction, fantasy and horror in Brazil 1875-1950), and Ramón López Castro's *Expedición a la ciencia ficción mexicana* (2003; An expedition to Mexican science fiction).

35. However, the overlap of genres has been noted by other critics. David Seed makes an argument for the "hybridity" of genres of science fiction in his 2011 book *Science Fiction: A Very Short Introduction* (1). Roger Luckhurst also remarks on the "hybridization and regenerative implosion of gothic, sf and fantasy" (420) in the 1990s British market in his 2003 article, "Cultural Governance, New Labour and the British SF Boom." Milner also points out that author China Miéville has made a case for the use of estrangement in fantasy and weird

fiction as in science fiction (106). In his *Terminal Identity* (1993), Scott Bukatman includes a fascinating discussion of cyborgs and the body, examining the crossover between science fiction and horror.

CHAPTER I

1. In her 2000 article, "Envisioning Cyborg Bodies: Notes from Current Research," Jennifer Gonzalez has examined eighteenth-century mechanical mistresses, modernist cyborg-like images of women, and women of color in the visual arts and in Japanese graphic novels (manga). This is a model for a diachronic approach of the female cyborg.

2. In *How We Became Posthuman: Virtual Bodies in Cybernetics, Literature, and Informatics* (1999), N. Katherine Hayles explores different iterations of the mind and body in the Western tradition. In her analysis of information science and its portrayal in science fiction texts, Hayles distinguishes three phases in cybernetic theory. The first is the homeostatic model, the second is characterized by "reflexivity," and the third phase of cybernetic theory is based on pattern recognition and randomness (35). These are discussed in detail throughout the chapter.

3. Stefan Gandler interprets Echeverría's baroque ethos by saying, "it does not contain any anti-capitalist tendency, but nevertheless entails the incessant attempt to break the rules of capitalist relations of production. . . . As an example, we could think here of the great festivities in which, even amid a situation of suffering and repression, participants find moments of undeniable happiness" (301).

4. Gandler notes this contradiction: "In the baroque *ethos* of capitalist modernity, we find a conflictual combination of conservatism and non-conformity" (301). He offers the example of the Zapatistas, who simultaneously embody tradition in their manner of dress and speech and a "rebellious and almost revolutionary pulse" (302).

5. J. Andrew Brown has addressed female cyborgs in his study *Cyborgs in Latin America* (2012); he finds more resistance to political oppression in experimental works of literature dating from 1982 to 1998 that involve a cybernetic presence. Several feature female cyborgs and address the aftermath of dictatorships in Argentina and Chile and the neoliberal regimes in works by Alice Borinsky and Eugenia Prado. Brown also analyzes Mexican author Carmen Boullosa whose *Cielos en la tierra* features an avatar called Lear who witnesses the apocalyptic savagery of a failed posthuman society (54–56). There are no parallels to this text in Brazil.

6. For a comparison between Hoffmann and Machado, see James R. Krause's "Enucleated Eyes," an examination of the father figure in the context of the Freudian definition of the uncanny and the castration complex in these texts

(58–62). Freud's view of the uncanny is fundamental to the interpretation of the body and horror, since it is based on the return of the psychic repressed that is triggered by physical sensations of fear, the double and unconscious repetition, and disorientation of the ego. See Freud's "The Uncanny," 211–12.

7. As Beckman explains, the commodity's "hidden abode of production" is separated from circulation and consumption because of a sophisticated division of labor that appears like magic on the market: "This illusion becomes even more compelling when we turn to the global reach of the commodity system, in which commodities traverse not only huge expanses of space and time, but also vastly different social formations" (32). Joseph Yanielli goes so far as to speculate that Marx may have come up with his idea of the commodity fetish from the antislavery movement that tried to expose the slave labor behind sugar consumption, which showed an image of a mutilated slave in 1787 and stated: "This is the price of the sugar you eat in Europe" in his article "Your Slavery Footprint."

8. As mentioned earlier, export reveries, blindness, and insanity are most effectively explored in *Quincas Borba* (1890), Machado's novel about the coffee boom, in which the protagonist, Rubião, has a name similar to that of the coffee plant *rubiaceae* found in subtropical regions like Brazil (Gledson 72). At a key moment in the novel, Rubião faints —in a temporary type of blindness—when faced with the punishment of a slave and later goes mad once he is swindled out of his fortune by his partner and his beautiful wife. Curiously, "O capitão Mendonça" was written the year that *Quincas Borba* takes place: 1870. While Amaral is able to resist and escapes unscathed, Rubião returns to Minas Gerais where he succumbs to madness and death.

9. Machado's gothic stories, published between 1870 and 1884 and collected in Magalhães Júnior's 1973 *Contos fantásticos*, predate what Alexander Meireles da Silva (2010) has characterized as the turn of the century "gothic science fiction" in authors such as Coelho Neto and João do Rio. Most literary critics have placed do Rio in the category of premodernist writer or as a literary chronicler primarily of interest for his documentation of Rio's street life and underworld. See M. V. Serra and Antonio Edmilson Martins Rodrigues's introduction to João do Rio's *A alma encantadora das ruas*.

10. Machado was an Afro-descendant who suffered from epilepsy and stuttering, while João do Rio was openly homosexual. Thus, both were vulnerable to the kind of pseudo-scientific discourse popularized under the term "degeneration." For more on this topic, see Dain Borges's 1993 study "'Puffy, Ugly, Slothful and Inert': Degeneration in Brazilian Social Thought 1880–1940."

11. Admirably recovered and analyzed by Haywood Ferreira for its connections to nineteenth-century theories of hypnotism and magnetism, the work pays homage to "Quaerens," the protagonist of the novel *Lumen* (1872), a work of science fiction by French astronomer and spiritist Camille Flammarion (Haywood Ferreira, *Emergence* 158).

12. For further details about the plot of *Querens*, see Haywood Ferreira (*Emergence* 157-162). While Haywood Ferreira focuses on the scientist and his connection to Frankenstein's hubris and failed scientific experimentation (162), Gabriel Trujillo Muñoz focuses on the theme of failed romance in Mexican science fiction (*Biografías* 52-53). More recently, in *Los espíritus de la ciencia ficción* (2017), Luis Cano examines Castera's text as one of the three foundational novels of Latin American science fiction, arguing that the spiritual "sciences" of the nineteenth century influenced the course that the genre would take in the region. He focuses on spiritism's insights into female psychology and the homosocial desire between the two male characters of *Querens* (204-215).

13. However, even with local industrialization, fetishism can occur and can affect everyday consumables, not only luxury goods. Often local traditional products are displaced by "new" valued-added ones. Gandler gives the example of white bread, which is often preferred by certain elites in Mexico over tortillas, since it is believed that white bread simply tastes "better" (250), illustrating how the industrial product is associated with modernity. Echeverría points out that, traditionally, the field of economics has emphasized that tools and industrial systems add value while generally ignoring labor, consumption, or the use-value of the product. For Echeverría, a product seeks to "communicate" or establish value through its purchase and consumption, not strictly its market value.

14. Hayles bases her analogies on Mark Seltzer's *Bodies and Machines* (1992). She summarizes that, "the first law of thermodynamics, stating that energy is neither created nor destroyed, points to a world in which no energy is lost. The second law, stating that entropy always tends to increase in a closed system, forecasts a universe that is constantly winding down. . . . The body is like a heat engine because it cycles energy into different forms and degrades it in the process; the body is not like a heat engine because it can use energy to repair itself and to reproduce" (*Posthuman* 101).

15. A female cyborg figures in a science fiction story by Frenchman Auguste Villiers de L'Isle Adam, *Tomorrow's Eve* (1886), where the helplessness of the artificial female body is emphasized, as in many Latin American tales. Villiers's novel, described often as a blend of the myth of Pygmalion and *Frankenstein*, involves Thomas Alva Edison, who invents an artificial female with an intellect to fulfill the desires of a rich baron. While the story has been linked to anxieties surrounding the medicalization of the female body and its supposed "aberrant" biology, it also undermines its "cyborg" nature by imbuing the creation with the "soul" of a living woman. See Tara Isabella Burton's article "Hadaly, the First Android." The extraction of female souls reminds us of "El donador de almas" (1899) by Amado Nervo, in which André, a neo-mystic, was able to harness female souls and place them in bodies. See Chapter 2 on gender for more about this story by Nervo.

16. Here I refer to Foucault's *Discipline and Punish*, which deals with the rise of prison systems and internal controls of the mind, and the *History of Sexuality*, on the control of the sexual body and the rise of the medicalization and informationalization of sexuality.

17. Hayles cites one of Wiener's contemporaries, Claude Shannon, who conceived of entropy as working with information in order to form a "thermodynamic motor" that would "drive systems into self-organization" (102).

18. Later, in *Autopoeisis and Cognition* (1980), Maturana and Varela theorize that the scientist/observer is part of the system, breaking down the notion of an objective reality that stands outside the observer and the observed (qtd. in Hayles 134–37).

19. For more about Queiroz's proto-feminism and this story, see Luciana Monteiro's 2013 article "Mulheres, mães e alienígenas."

20. The figure of the gendered cyborg does not emerge again until the first decade of re-democratization, when science fiction writer Cid Fernandez gives his female cyborg protagonist the capability and desire to reproduce in a novelette called "Julgamentos" (1993; Judgments). Published nearly twenty years after Abreu's text, Fernandez's heteronormative female cyborg wishes to marry and have children with her cyborg husband. See Ginway, "The Body in Brazilian Science Fiction" (202-5). A similar desire to reproduce is expressed by robots in Jacques Barcia's 2009 steampunk story "Uma vida possível atrás das barricadas," see Ginway, "Posthumans in Brazilian Cyberpunk and Steampunk" (244).

21. Artistic forms of resistance and cooptation of Afro-Brazilian practices such as capoeira and samba are the clearest examples. Legalized by the Vargas regime, they soon became products that would become nationalized and exported and used to promote Brazilian "racial democracy." See Stanley Bailey, *Legacies of Race* (2009) and Bryan McCann, *Hello, Hello Brazil: Popular Music and the Making of Modern Brazil* (2004).

22. Among the most visible of this type of crime is the shooting death of Rio councilwoman Marielle Franco in March of 2018 and the resignation of a gay rights advocate and congressman Jean Wyllys. Both events illustrate the level of violence against the LBGTQ population, not to mention the anti-gay rhetoric of Brazil's Jair Bolsonaro. See José Miguel Vivanco, "A Voice for LBGT Rights Silenced in Brazil" (Human Rights Watch, January 24, 2019).

23. Tragically, Caio Fernando Abreu would fall victim to this disease some twenty years later, dying in 1996; see Arenas 13. The AIDS epidemic is generally considered to have begun in 1981; therefore, Abreu could not have known about it while writing his story in 1975.

24. For Süssekind, the function of the *romance-reportagem* was to portray reality that had been censored in the media: such novels would "dizer o que impedia o jornal de dizer, fazendo em livro as reportagens proibidas nos meios de comunicação de massa; a produzir ficcionalmente identidades lá onde dominam as divisões, *criando uma utopia de nação* e outra de sujeito, capazes de atenuar a experiência cotidiana da contradição e da fratura. Para exercer tais funções a literatura opta por negar-se enquanto ficção e afirmar-se como verdade. O naturalismo torna-se todo-poderoso" (57, emphasis added; say what

the newspaper would not, putting all the censored reports in mass media into a book; to fictionally produce identities where divisions dominate, creating one national utopia and another of subjects, capable of softening the daily experience of contradiction and rupture. In order to carry out these functions, literature chooses to negate itself as fiction and affirm itself as truth. Naturalism becomes all powerful.) This illustrates the power of such "realist/naturalist" texts, which displaced other forms of literature in national discourse.

25. Ironically, the lack of reform of the police and security operations, which incorporated death squads and other extra-legal killers, created a vacuum for illegal networks to flourish. The development of a dual state, in which rights of citizenship and access to state institutions are divided, is part of this legacy. Extra-legal militias, as well as organized crime, continue to grow in the aftermath of dictatorship and the introduction of neoliberal policies. See Ben Penglase, *Living with Insecurity in a Brazilian Favela* and Desmond Arias, "The Impacts of Differential Armed Dominance of Politics."

26. Little criticism exists on González. In his article on "Rudisbroeck o los autómatas" (1978), Serrato Córdova focuses more on the gothic elements, paying special attention to the intercalated theater production in the story, which is followed by a Grand Guignol–style decapitation and a freak show. He makes only passing reference to the automatons.

27. Ignacio Padilla's award-winning 2003 story, "Las furias de Menlo Park" (The furies of Menlo Park) appears in the collection *Los androides y las quimeras* (2009; Androids and chimeras). As an author associated with Mexico's so-called "Crack" generation, which protested the tenet that Mexican literature should always be about Mexico and argued for a more global perspective, Padilla sets his story in the United States and uses the cyborg/android to question gender roles and violence against women. Although Padilla sets his story in the US, far from the Mexican border, it still recalls the murders of female workers in the *maquiladoras* (large-scale factories) associated with border towns and especially Ciudad Juárez. In Padilla's story, the loneliness and isolation of female workers is captured by the fragile cyborg/doll figure whose story is finally "heard," albeit indirectly, through the memories of the protagonist, who had planned to reveal the truth about Edison's illicit affairs and the suicide of one of his lovers upon the death of the famous inventor. Both stories by González and Padilla have elegiac qualities that mourn and resist modernity in a way typical of the baroque ethos.

28. The idea of time that passes without generating "history" is captured by Mexican Emiliano González in a 1978 story called "La herencia de Cthulhu" (The heritage of Cthulhu), a reference to the atavistic deity created by the American master of horror, H. P. Lovecraft. A subsection of the story refers to the "Museo del Chopo" in Mexico City, and its Museum of National History where, until 1964, circus and sideshow remnants of the past constituted a baroque museum of bodies, books, and substances whose collective effect was to provide an alternative view of the positivist ideas of progress and science.

29. The interplay of presence, absence, and supplementarity are described in Derrida's "Structure, Sign and Play" (260–61).

30. The focus on organized crime and drug trafficking shifts the blame onto displaced members of society and, in the case of Brazil, usually young men of color, criminalizing poverty, while hiding those that benefit from the drug trade. Brazil's endemic lack of social mobility combined with drug trafficking led to widespread violence during the 1980s and 1990s. This violence, which affected primarily young men of color, is portrayed in the film *City of God* (2002), directed by Fernando Meirelles and Kátia Lund. In an article about the role of violence in the Latin American cinema of the 2000s, Ignacio Sánchez Prado uses *City of God* and Mexican filmmaker Alejandro González Iñárritu's 2000 *Amores Perros* to illustrate the ideology behind "exoticized violence" during the period of neoliberalism: "In recent years, this destabilized culture has produced new images of violence that allegorize the sense of uncertainty, which is a product of the fall of the paternalistic state," along with a new "citizenship of fear" (*"Amores"* 39). As Sánchez Prado observes, this type of cinematic violence decontextualizes the political implications of neoliberal policies, deepening the rift between the people and their government that began in the 1990s with free trade agreements. Privatization and the dismantling of the state, initially undertaken in order to protect local industries and workers, left a political power vacuum and economic instability and unemployment in its wake.

31. In a study on the emergence of global markets and neoliberal polices from 2002, Jean and John Comaroff draw parallels between the sudden accumulation of wealth via financial dealings and that achieved by the underground economies of organized crime, since neither relies on ordinary labor costs or infrastructure (786). Similarly, for Beckman, the sudden accumulation and evaporation of wealth recurs in the stock market with global repercussions, illustrating how "instability, mobility and extreme inequality" (xviii), characterize capital and commodity speculation in Latin America.

32. These two cyberpunk texts have already received ample critical coverage. However, I would also emphasize the war/espionage/thriller aspects of *La primeira calle de la soledad*, since its protagonist often depends more on his physical skills as a fighter than on his mental skills as a hacker. Calado's story offers a futuristic extrapolation of Rio's crime networks, and refers to one group by name, the Comando Vermelho (Red Command), which is based on an actual crime ring that arose in the prison system under the dictatorship among leftist political prisoners whose ideology influenced inmates and whose networks began to include elements of civil society. See articles by Hernán García and Muñoz Zapata on *La primeira calle de la soledad* and Ginway on "O altar dos nossos corações" regarding the connection between narco-trafficking and cyberpunk in "The Body in Brazilian Science Fiction."

33. In *Powers of Horror: An Essay on Abjection* (1982), Julia Kristeva explains the corporeal reality of horror and how it is provoked by the breakdown of the distinction between the sense of self and other, especially as triggered by bodily functions (1–2). These sensations are also extrapolated to social taboos regarding hygiene and other associations with a "disowned" self (2–4).

34. Although Pepe Rojo's 1996 "Ruido Gris" (Gray noise) does not deal with drugs per se, it does foreshadow the public's addiction to news. Rojo's story describes the life of a journalist with an ocular implant who is employed as the ultimate eyewitness reporter: he films stories for the company that "owns" his implant merely by visually witnessing the events. Traumatized by his work, he fantasizes about self-mutilation and suicide while narrating incidents of violence, including a violent police raid of a kidnapping ring, terrorists who set off a bomb in a department store, and a series of suicides. Like Porcayo's *La primera calle de la soledad*, the story has a circular structure, beginning with the narrator's first assigned story, a person's suicide, and ending with his contemplation of taking his own life.

35. Gloria Cubil perceives nothing about her body. Sara Anne Potter convincingly proposes the disassociation of mind and body in Marshall McLuhan's *Understanding Media* (1964) as a way of understanding *Gel azul*, in "auto-amputation" or withdrawal from human connection in favor of technological pleasure" (Potter 151–52).

36. In *Gel azul*, despite his shock of becoming an amputee, Salgado explains the conspiracy: when the US legalized drugs, the world economy collapsed and a worldwide war broke out, thereby increasing the demand for prosthetic limbs. The war is being fought to support the global economy, not to combat powerful drugs, which are now used in harvesting limbs from unwary enemies. For more on this theme, see Robert Noffsinger, "The Body as Commodity: Posthuman Tendencies in Bef's *Gel azul*."

37. Cuenca's novel has been translated into English as *The Only Happy Ending to a Love Story is an Accident*, translated by Elizabeth Lowe (Dartmouth, MA: Tagus, 2013). As an upcoming young author with a marketable book, Cuenca is a mainstream writer who has ventured into cybernetic fiction.

38. The repeated text leading up to the accident is "No topo de tudo, um grande outdoor anuncia sopa em tubos de neon. O único conjunto de janelas sem cortinas fechadas ou vidros escurecidos é o do quinto andar do edifício curvo à direita. Ali, um grupo de pequenas bailarinas ensaia uma coreografia no centro da sala, enquanto outras alongam as pernas numa barra de metal" (17, 54, 106–7, 141–42; Above everything, a huge billboard announces soup in neon tubes. The only set of windows without closed curtains or darkened panes is on the fifth floor of the curved building on the right. There, a group of young ballerinas rehearse their choreography in the center of the room, while others extend their legs on a metal barre [My translation]).

CHAPTER 2

1. Nemi Neto analyzes realist texts or those that follow a biographical or autobio-
    graphical tendency: his literary corpus includes works by Caio Fernando Abreu,
    André de Figueiredo, and Daniel Herbert as well as several films. For a study on
    Latin American gay autobiography, see also Paola Arboleda-Ríos's 2013 Univer-
    sity of Florida dissertation "Homotramas: Actos de resistencia, actos de amor."

2. In his 1971 essay "O entre-lugar do discurso latino-americano" (The space in-
    between of Latin American discourse), Santiago writes: "Entre o sacrifício e o
    jogo, entre a prisão e a transgressão, entre a submissão ao código e a agressão,
    entre a obediência e a rebelião, entre a assimilação e a expressão—ali, nesse
    lugar aparentemente vazio, seu tempo e seu lugar de clandestinidade, ali, se
    realiza o ritual antropófago da literatura latino-americana" (28; "Between
    sacrifice and play, between prison and transgression, between submission to
    the code and aggression, between obedience and rebellion, between assimi-
    lation and expression—there, in that apparently empty place, its time and its
    place of hiding, the anthropophagic ritual of Latin American literature takes
    place"). Thus, belonging neither to cultural center or periphery, the place of
    the Latin American author is always critically located in a "space in-between."

3. Nemi Neto studies literature and film that focus on male subjects, although he
    does mention the history of the lesbian movement in Brazil. He also includes
    Tarsila do Amaral's 1928 painting *Abaporu* (Tupi for "The man who eats") to
    illustrate the centrality of the body over the mind in Oswald de Andrade's
    theory of antropofagia. Notably, the ambiguously sexed indigenous person in
    Tarsila's painting, rendered in a primitivist style with a tiny head and a giant
    foot in the foreground, suggests the opposite of the privilege of the head in
    Western culture. The omission of female writers and artists in Neto's corpus
    may be filled in part by Andrea Villa's University of Florida dissertation "Con-
    trapunteo neobarroco" (2015), in which she suggests that the use of absence
    and empty space in the artistic works of Doris Salcedo and Lygia Clark may
    be an expression of the female neo-baroque, suggestive of Santiago's space
    in-between.

4. Chapter 1 of Nemi Neto's dissertation, "The Anthropophagic Queer," offers
    an overview of gay and lesbian studies in Brazil, a field that has been covered
    by anthropological history or cultural studies and has not been as widely
    addressed in literary studies (28).

5. Sexuality and science fiction begin to appear in Mexico with Arturo Rojas
    Hernández's neo-baroque homosexual space opera *Xxÿëröddny, donde el gran
    sueño se enraíza* (Xxÿëröddny, where the great dream takes root; written under
    the pseudonym Kalar Sailendra). Mario Bellatín's elegy to the AIDS epidemic,
    the dystopian *Salón de belleza* (1994; Beauty salon), has queer themes as does H.
    Pascal's interstellar pirate novel *Fuego para los dioses* (1998; Fire for the gods),

which also has trans characters. Male authors writing from a female point of view include Bef (Bernardo Fernandez) in his 1996 story "Las últimas horas de los últimos días," (The last hours of the last days) and Omar Avilés in "El pantano de los peces esqueleto" (2010, The swamp of the skeleton fish), both of whom offer flexible gender roles to their male and female characters. Enrique Serna's story "Tía Nela" (Aunt Nela)" about an anguished trans character, appears in his collection of science fiction stories *El orgasmógrafo* (2010).

In Brazil, Lygia Fagundes Telles's story "Tigrela" (1977), as Anne Connor has noted, is a story of the "feline fatale" in Latin America, and fits into the "cat woman" theme of queer sexuality. André Carneiro's story "Transplante de cérebro" (1978; "Brain Transplant" [2003]), written during the first period of political opening of the military dictatorship, is rich in its exploration of sexuality from multiple viewpoints, featuring multi-gendered characters (Ginway "Transgendering"). In addition, several authors write from the point of view of a different sex, including Finisia Fideli's "Quando é precisa ser homem" (1993; Sometimes you gotta be a man) as well as Gerson Lodi-Ribeiro who adopted the pseudonym Carla Cristina Pereira to write alternate histories such as "Xochiquetzal" (1999) and "Longa viagem para casa" (2000; The long voyage home). He writes about his experience in the essay "Os anos em que fui Carla" (The years I was Carla), noting that as Carla he won prizes and received propositions.

For more on women science fiction and fantasy writers, see Ginway's 2007 "Recent Brazilian Science Fiction Written by Women" and Gabriela Damián Miravete's "La mano izquierda de la ciencia ficción mexicana."

6. An anthology of stories exclusively by women is *La imaginación: La loca de la casa* (2015, The imagination: The madwoman of the house), ed. by Bef.

7. The focus of Wendy Gay Pearson's introduction, "Alien Cryptographies," to *Queer Universes: Sexualities and Science Fiction* (2008) confirms the technological and alien presence involved in alternate sexuality in American science fiction.

8. Jáuregui states that Andrade's cannibal is "reformulado como el ser liberado del trabajo, de los miedos metafísicos y de las restricciones autoritarias gracias a la tecnología, la recuperación de la alteridad, el fin del mesianismo, la muerte de Dios y el reemplazo del Estado por el matriarcado de Pindorama" (582; later reformulated as a being liberated from work, metaphysical fears and authoritarian restrictions thanks to technology, [and] the recuperation of alterity, the end of messianism, the death of God and the replacement of the state by the matriarchy of Pindorama). Jáuregui goes on to note that there is a postwar faith in technology, which, paradoxically, Andrade viewed as the end of Western rationality, elements that appear in Andrade's 1953 series of essays collected under "A marcha das utopias" in *Do Pau-Brasil à antropofagia e às utopias* (145–228).

9. For Roberto Causo, *A Amazônia misteriosa* actually surpasses H. G. Wells's *The Island of Dr. Moreau* (1899) in importance, because it includes an anticolonial perspective and avoids religious moralizing (*Ficção* 176–77).

10. To be fair, Del Picchia's *A filha do Inca* (1930) also mentions Incan culture, but mostly to the extent that the Incan princess provides a love interest for the Brazilian hero. See Ginway, "The Amazon in Brazilian Speculative Fiction" (3).

11. The German succeeded in reversing the aging of a sixty-year-old woman by implanting the ova of a spider monkey into her. The woman now appears some thirty years younger and is taking care of her interspecies baby, whom she is teaching to speak. His interspecies experiments recall an early story by Gustave Flaubert, "Quidquid Volueris" (1837), in which a French man goes to Brazil and forces a female slave to mate with an ape, creating a hybrid interspecies of human and animal. In an insightful article dealing with the human/ape interface, Joan Gordon uses the term "amborg" for the interspecies offspring. In Flaubert's story, the mother dies and the son is taken back to Europe, where he eventually kills himself. Gordon reads this as the actions of a Byronic hero (256), since the son's inability to speak contributes to his violent actions. While Cruls endows his amborg with less subjectivity, the child's ability to learn to speak may suggest a different fate. In Cruls's novel, both the Amazon women and the hybrid children cross genders and species, thus questioning the parameters of identity and humanness. Roberto Causo speculates that the novel would have improved had Cruls focused on deepening this theme rather than focusing on the adventure plot and local color (*Ficção* 176).

12. See John Rieder's discussion of the "lost-race motif" of the late nineteenth and early twentieth century to understand the formulaic nature of such narratives (22).

13. In a sense, Cruls description of Hartmann's research anticipates the medical experiments of the German Reich, which included artificial and cross-species insemination. (Peter Tyson, "The Experiments," *Holocaust on Trial.* Nova Online, Oct. 2000. www.pbs.org/wgbh/nova/holocaust/experiside.html.)

14. The queer is also present in Mário de Andrade's modernist classic, *Macunaíma* (1928), a text that also evokes the myth of the Amazon women warriors. In his groundbreaking work *The Fantastic: A Structural Approach to a Literary Genre* (1975) Todorov's concept of the "hyperbolic" or "exotic marvelous" (54–55) explains Macunaíma's transformations, because supernatural powers and beings transform reality in a way that is accepted by all characters, its trickster hero Macunaíma is an aggregate of Brazil's founding cultures (African, Amerindian. and European) who is capable of magically changing race, class, and gender. The basic plot consists of Macunaíma's attempts to recover an amulet given to him by his former lover Ci, the Queen of the Icamiabas, the original Amazon women. In a "fantasy of reverse colonization," (Ginway, "Teaching Latin American Science Fiction" 185) those who are less powerful in society—

indigenous peoples, Afro-descendants and women—have their revenge against the more powerful, a feature often seen in Latin American science fiction. As author and critic João Silvério Trevisan has commented, Mário de Andrade's family and subsequent scholarship have attempted to repress knowledge of his homosexuality (256). As a man of mixed race, a musician, and scholar who also happened to be a homosexual, Mário de Andrade had a vision of race and sexuality that was far beyond that of his peers.

15. La Malinche or Doña Malina was a Mayan woman and translator for Hernán Cortés as well as the mother of his first son. She remains is a potent icon in Mexican culture to this day, as any person who "sells out" to outside interests can be accused of *malinchismo*. See Pirott-Quintero for further discussion of the issue of origins and Mexico's foundational myth of Malinche in *Duerme*. Jean Franco's "La Manliche: From Gift to Sexual Contract" (1999) is a more nuanced study on this figure.

16. As Rachel Haywood Ferreira notes, the reaction to the Darwin's ideas of evolution took on a different meaning in Latin America, where Larmarckian eugenics and the inheritance of acquired characteristics overlapped with the idea of the perfectability of the soul through the alternative sciences of spiritism, Theosophy, and mesmerism (80), a tendency also mentioned by Luis Cano in his study *Los espíritos de la ciencia ficción*, which includes an analysis of *Querens* (190–95). This influence is present in the works of Mexican *modernistas* Castera and Nervo as well as Brazil's Coelho Neto.

17. For more conventional studies on *El donador de almas* and studies on Mexican modernism see Karen Poe, *Eros pervertido: La novela decadente en el modernismo americano*, and Robert McKee Irwin, "Lo que comparte el positivismo con el modernismo mexicano: El hemafroditismo, la bestialidad y la necrofilia."

18. Rafael exclaims the following upon Alda's entry into his brain: "Tener en sí a la amada, en sí poseerla. ¡Acariciarla acariciándose! Sonreírla, sonriéndose . . . glorificarla, glorificándose. . . . Cierto, algunas veces tales y cuales miserias fisiológicas ruborizaban al doctor por ministerio de su semi-cerebro" (41; To have his beloved within, to possess her within. To caress her, by caressing himself. To smile at her, smiling at himself . . . to glorify her, by glorifying himself. . . . True, from time to time certain physiological necessities made the doctor blush because of the presence in the other half of his brain).

19. Rafael vents his anger, criticizing him thus: "Poeta desequilibrado, romanista, esteta, simbolista, ocultista, neomístico o lo que seas" (48; Unbalanced poet, romance philologist, aesthete, symbolist, occulist, neomystic, or whatever you are).

20. Since there are few studies on embodiment and shifting gender identities in Latin American literature, I cite here Elizabeth Grosz's 1994 *Volatile Bodies: Toward a Corporeal Feminism*, which offers an interdisciplinary approach to the history of the body in the fields of psychology and philosophy. Building on concepts of the body developed by French feminists and European phi-

losophers, Grosz comes closest to describing a state between genders. Using the traditional association of female bodies with "fluidity" as opposed to the "solidity" of male bodies (202-5), she captures the sense of flux or slippage between these two poles that defines the queer body and ambiguous gender identity, recalling another meaning of Santiago's space in-between.

21. See studies by Arata and Moretti for postcolonial and Marxian analysis of *Dracula*.

22. As a side note, in the article "Lesbianas en América Latina: de la inexistencia a la visibilidad," Sardà, Posa Guinea, and Villaba emphasize not only the taboo of lesbianism, but the silence and invisibility of lesbians' struggle. Also interesting is their observation that, in contemporary Mexico and Brazil, lesbians have separate pride marches to advocate for their rights (9).

23. While several Brazilian authors of this period explore the theme of homosexual or queer identity at the turn of the century, they tend to write in the realist/naturalist style. For example, Aluísio Azevedo portrays an openly gay male who washes clothes beside the other laundresses as well as alongside lesbian prostitutes in *O cortiço* (1890; The tenement), Adolfo Caminha's novel *Bom-Crioulo* (1895; The good black man) recounts the passion of a black sailor for a blond cabin boy, and Domingos Olímpio's novel *Luzia-Homem* (1902; Luzia-man) features a masculinized female protagonist of impossible physical strength. Finally, João do Rio's short story "História de gente alegre" (1910; A tale of happy people) explores a death resulting from a lesbian encounter.

24. Alexander Moreira-Almeida et al. cites concerns by physicians regarding madness and spiritism in Brazil, noting that in 1909 a conference was held at the Medical Society of Rio de Janeiro called The Dangers of Spiritism, but no action was taken against the movement (9). This conference took place one year after the original publication of *Esfinge*.

25. In Mexico, national consolidation and the strengthening of the political body continued with the institutionalization of the Partido Revolucionario Instituticional, or PRI, an incarnation of the corporate state. This socialist vision of a strong centralized state can be paralleled with Francisco L. Urquizo's *Mi tío Juan: Novela fantástica* (1934), which features a male cyborg.

In Brazil, there are examples of atypical male and female characters in del Picchia's triology of adventure novels for juvenile audiences the 1930s—*A filha do inca* (1930), *Kalum* (1936), and *Cummanká* (1938)—in which technology is associated with cyborgs, gynocracies, or imperialist cultures that a masculinist, sovereign Brazil must defeat. See Ginway, "The Amazon" (3-5), regarding its science fiction roots, and Queluz and Queluz, "*Kalum*: Utopia e modernismo conservador," for its links to conservative political thought and eugenics in Brazil.

26. Tiptree's "The Women Men Don't See" has a famous line which the male narrator learns from one of the women that "We survive in ones and twos in the chinks of your world machine. . . . We are used to aliens" (142).

27. The anxiety to rebuild Mexican society after the Revolution appears in the eugenic theme of Eduardo Urzaiz's 1919 *Eugenia*. As Haywood Ferreira has shown, the work advocates for several feminist causes, including the education of women and the elimination of sexual double standards in traditional marriage, given that women are no longer expected to give birth and raise children and are free to pursue intellectual activities or other work (*Emergence* 67-79). Men become the incubators of human offspring. At the same time, the novel questions the benefits of these social changes in its portrayal of Celiana, the main character. While she enjoys the psychological and financial independence afforded by her position as a lecturer in sociology, she also suffers from depression. While Rebolledo's Elena destroys the men around her, Urzaiz's Celiana slowly destroys herself, becoming depressed after her younger lover, Ernesto, rejects her, smoking cannabis to the point of becoming "un cadáver galavanizado por la esperanza" (118; a cadaver held together by hope). The original 1919 image on the cover of *Eugenia* (Haywood Ferreira, *Emergence* 71) portrays Celiana as thin and angular, with small breasts that make her appear alternately seductive, androgynous, and vampire-like. Her phallic object is the cigarette she holds in her hand, whose plumes of smoke suggest the snake-like hair of Medusa and the hair of the figure upon which she rests her arm, which is the profile of a dark full-breasted female figure whose face is a skull. Celiana provides the problematic and powerful side of the phallic female figure, which lends complexity to this utopia, because Celiana, the novel's most sympathetic character, is destroyed by its eugenic ideals.

28. For the evil attributed to the salamander see Marty Crump, *Eye of Newt and Toe of Frog, Adder's Fork and Lizard's Leg: The Lore and Mythology of Amphibians and Reptiles*, 147.

29. Tarazona is not the only author to speculate about oviparity. See Ginway, "Simios, ciborgues y reptiles" (2017), in which I compare her work to stories by Brazilian author Finisia Fideli and Puerto Rican author Marta Aponte Alsina.

CHAPTER 3

1. For reasons of clarity, I will reserve the term undead to designate vampires, since only the latter can be described as fully alive, or at least far more alive than zombies. Note also that zombies and vampires differ from the truly dead, who still manage to narrate their stories in works such as Machado de Assis's *Memórias póstumas de Brás Cubas* (1881; *The Posthumous Memoirs of Brás Cubas*), whose protagonist recounts the events of his life with the narrative distance that only death can provide. The same holds true for stories such as Maria Elvira Bermúdez's "Soliloquio de un muerto" (1951; Soliloquy of a dead man), a philosophical meditation "de ultra tumba," and Juan Vicente

Ripoll Melo's "Mi velorio" (1956; My wake), which, like *Brás Cubas*, is narrated from beyond the grave. Such narratives tend to be commentaries on the hypocrisy of the living.

2. For the popularity of such liberal doctrines in the late nineteenth century, see Joseph Love, *Crafting the Third World* (1996), 395.

3. Ericka Beckman asserts that between 1870 and 1930 the Export Age experienced financial instability due to market integration and liberalization, which led to cycles of boom and bust throughout Latin America (83-85). Beckman cites the 1890 Baring Crisis in Argentina and the 1891 *Encilhamento* (saddling-up) stock-market crisis in Brazil (95) as two prime examples of this instability, noting that in Argentina an entire subgenre of novels about the stock-market crash emerged between 1890 and 1900 (86).

4. Afonso Henrique Lima Barreto (1881–1922) was born into a humble family and suffered racial persecution at the hands of one of his teachers in college. He abandoned his studies when his father was interned in a mental institution in 1902. Although he suffered from alcoholism, Lima Barreto had a prolific career as an author and journalist before he died tragically at age forty-one. He is considered to be a precursor to modernist writers for his use of irony and straightforward language. His works range from the parody of Brazilian nationalism in *Triste fim de Policarpo Quaresma* (1915; The sad demise of Policarpo Quaresma) to his description of his own internment in a mental institution, *Diário do hospício e Cemitério dos vivos* (A hospital diary and Cemetery of the living) published posthumously in 1953. See Antonio Arnoni Prado, *Lima Barreto: Literatura comentada*.

5. Another useful approach to literary zombies is Terry Harpold's "The End Begins: Wyndham's Zombie Cozy," in which he relates the sudden appearance and unsteady gait of the triffids in John Wyndham's *The Day of the Triffids* (1952) to the perception of the uneven pace of economic change and "unfinished history" (160). Zombie stories often contain strange time lapses and fissures, through which the zombie trope captures the inarticulate anxiety that accompanies the incursion of social and technological change. Harpold notes the alternation of speeds in triffid behavior: their slow, unsteady, relentless advance followed by the whip-like speed of their stingers, then again the slow feeding on the dead. This uncanny change of pace becomes a metaphor for human reactions to new cultural and economic paradigms, which shift and lurch, making the new order impossible to predict. For Harpold, this is the "unmeasured interruption of history's seriality" (161)—or the state of permanent "stasis/crisis" of a zombie invasion.

6. For more on Brazilian proto-zombies see Ginway, "Eating the Past: Proto-Zombies in Brazilian Fiction 1900–1955" (2018). This article includes analyses of Murilo Rubião's "O pirotécnico Zacharias" (1943; Zachariah, the pyrotechnician), Gilberto Freyre's 1955 retelling of the legend of "Boca-de-ouro" (Mouth-of-gold) and Roberta Cirne's recent graphic novel based Freyre's tale. All of these stories feature the zombies displaced in time and space that haunt modern Brazil.

7. Uprisings continued to surface in Brazil at various moments, including Brazil's 1897 War of Canudos in the state of Bahia, the 1904 revolt against mass vaccination, and the 1910 protest by sailors against the use of corporal punishment, to name a few. By the 1920s, the middle sectors of society would join in protests; witness the march of soldiers through the Brazilian interior (A Coluna Prestes [The Prestes Column]), led by future communist leader Carlos Prestes (1924–1927), as well as other marches led by other liberal officers of the armed forces in the Movimento Tenentista (The Lieutenant Movement; 1924). See Boris Fausto, *História concisa do Brasil*, 166–76.

8. Curiously, the scenario resembles that of Marcelo Paiva's 1986 novel *Blecaute* (Blackout) published shortly after the end of the dictatorship in Brazil. Both novels trace the story of young people who live in a world in which an unexplained disease has killed the majority of the population. Paiva's novel explores the lives of middle-class survivors and ends on a note of hope when a couple has a child, a symbol of renewal.

9. Gender is not often a consideration in zombie narratives, yet Aridjis's concern with human trafficking brings it to the forefront. The novel explicitly refers to the connection between narco-trafficking and violence against women: "La narco-guerra había incrementado la violencia contra las mujeres. El crimen organizado tenía muchos rostros, entre ellos el del feminicidio. La inacción de las autoridades era una forma de complicidad y ocultaban las desapariciones" (30; The narco-war had increased violence against women. Organized crime had many faces, among them femicide. The authorities' inaction was a form of complicity and they covered up the disappearances).

10. See Jessica Loudis for more details regarding cartels: "El Chapo: What the Rise and Fall of the Kingpin Reveals about the War on Drugs" *Guardian*, June 7, 2019.

11. Haroldo de Campos points out that this was also the strategy of Brazil's baroque poet Gregório de Matos (1636-1696), which makes him, in Campos's view, the precursor to Andrade as a poet/cannibal (16).

12. The content of this chapter was inspired by Sherryl Vint's essay in *Zombie Theory*, "Abject Posthumanism: Neoliberalism, Biopolitics and Zombies," which includes references to Latin American zombies and Agamben's "In Praise of Profanation."

13. According to Elio Gaspari's study *A ditadura escancarada* (2002; The blatant dictatorship), the number of complaints of torture averaged about one thousand per year from 1969 to 1972. During the same period, opposition guerrillas killed some sixty people (470–74).

14. Veríssimo's saga begins in the mid-nineteenth century, narrating the story of two rival landowning families, the Vacarianos and the Campolargos, who establish their power after the Paraguayan War in 1870. The narrative leads readers through the political regimes from the reign of Dom Pedro II to the declaration of the Republic in 1889, and finally to the oligarchies that ruled

until 1930, when Getúlio Vargas, Rio Grande do Sul's most important politician, came to power. Vargas's party and his followers (João Goulart and Leonel Brizola) largely determined the course of Brazilian politics leading up to the military coup in 1964.

15. Few books were officially censored, since most censors were more concerned with magazines, newspapers, and television, as mentioned by Nancy Baden (110). Although Baden examines novels that criticize the regime, including Veríssimo's *Incidente em Antares*, she does not mention the torture scene depicted in the novel.

16. Swarms generally indicate oppression. In José J. Veiga's 1966 *A hora dos ruminantes* (trans. *The Three Trials of Manirema*), cows imprison the inhabitants of a small town to illustrate their subjugation by authorities, while in Miguel Jorge's 1980 "Véspera de pânico" (The eve of panic), it is invading grasshoppers that terrorize residents. In Mexico, José Luis Zárate uses albino ants to symbolize economic invasion from the United States in his 1994 novel *Xanto: Novelucha libre* (Xanto: A novel *lucha libre*). The swarm is a harbinger of destruction, but ironically in the story no one seems to notice or take action against it: "Hormigas albinas que empezaron a fluir como invisibles corrientes desde lo más alto del edificio: un mar de seres . . . que fueron a distribuirse por toda la ciudad. Casi invisibles, columnas terriblemente blancas que caminaban, sin fin alguno, incesantemente. . . . Y la ciudad se hacía la desentendida, como si esa invasión no tuviera nada que ver con ella." (57; White albino ants flowed like invisible currents from the high point of the building: a sea of beings . . . went about spreading themselves throughout the city. Almost invisible, their white columns marched, without any purpose, incessantly. . . . And the city paid no attention, as if this invasion had nothing to do with them.)

Zárate appears to imply that these insidious, seemingly invisible albino zombie ants threaten to turn Mexico's population into zombie workers. Finally, Karen Chacek's 2010 "La hora de las luciérnagas" (The hour of the fireflies) is about a deadly strain of fireflies that have a clear association with the holocaust and also with neoliberal zombified office workers.

17. For more details about this story as a zombie narrative, see James R. Krause, "Undermining Authoritarianism: Retrofitting the Zombie in 'Seminário dos ratos' by Lygia Fagundes Telles" in *Alambique* (2018). Krause associates the rats with both the exploited working classes and the Brazilian military, since "the image of the rat occupies different fields of significance depending on the perspective of the characters, the author, or the reader" (11). In my opinion, while the military leaders can be seen as soulless, they are not the resistant zombies represented by the rats.

18. The character may also be an allusion to the science fiction author and comic book artist Bernardo Fernández or Bef.

CHAPTER 4

1. For more on "A vida eterna" see Ginway, "Machado's Tales of the Fantastic: Allegory and the Macabre" (215-17).

2. There are many examples of the vampire in Mexico and Brazil during the 1965 to 1995 period. In Brazil, *O vampiro de Curitiba* (1965; The vampire of Curitiba) by Brazilian Dalton Trevisan, includes vampiric men who prey on women as Brazil industrializes and men lose their sense of patriarchal power in the urban setting. For a summary of his works, see Andrew Gordus's 1998 article, "The Vampiric and the Urban Space in Dalton Trevisan's *O vampiro de Curitiba.*" Some twenty years later, Marcio Souza uses vampiric imagery in his 1983 novel *A ordem do dia* (*The Order of the Day*) to criticize the Brazilian military government's policies of exploiting the resources of the Amazon during the regime's hold on power from 1964 to 1985. In Mexico, Cardona Peña writes about vampires invading Mexico in his story "Los hemoglobitas" (1966; The hemoglobites) in an allegory about Pemex, the Mexican oil company, and Lazlo Moussong's *Castillos en la letra* (1986; Castles in the letter) describes vampires as tourists in search of exotic blood. However, it is not until the 1990s, with Emilio Pacheco's "No perdura" (It will not last) from *La sangre de Medusa* (1990; The blood of Medusa) and Gabriela Rábago Palafox's *La voz de la sangre* (1990; The voice of blood), that authors use vampires as a way to explore the social body as a metaphor of parasitism and decadence, exposing the erosion of the institutionalized power of traditional families and the Catholic Church. For an analysis of diverse iterations of post-1990 vampires in Mexico, see José Luis Martínez Morales's 2012 article, "¿Vampiros mexicanos o vampiros en México?"

3. As Campbell explains, Esposito relies on Simone Weil's concept of the "impersonal" as that which is sacred and paradoxically shared by all humans, in a way that goes beyond the individual: "Citing Weil in this instance is intended to create conditions in which a radical opening of the human community to the impersonal can occur" (78).

4. In the essay "Taking Dracula's Pulse: Historicizing the Vampire," Lisa Lampert-Weissig has traced the eighteenth-century origins of the vampire myth in Eastern Europe, illustrating how it spread through Western Europe The collection *The Vampire Goes to College*, edited by Lisa A. Nevárez, offers a wide variety of approaches to teaching vampire literature.

5. The 1997 Norton edition of *Dracula* includes essays that offer diverse interpretations of Bram Stoker's work, from sexuality to feminism. See Christopher Craft for gender ambiguity and Talia Schaffer for homoerotic analysis.

6. Eljaiek-Rodríguez recovers lesser-known texts from Venezuela and Colombia in order to claim them as part of the tropical gothic alongside better-known horror tales by the likes of Gorriti, Quiroga, Holmberg, and Lugones, while also illustrating the social critique of horror in recent films from Argentina, Colombia, and Mexico.

7. A reading of the vampire as an allegory of colonial invaders is most evident in Méndez's *El vampiro*, as developed in Carmen Serrano's article "Revamping Dracula on the Mexican Silver Screen," where a French vampire is associated with the French invasion of Mexico and the rule of Maximillian (1864–1867). It is significant that the young female protagonist— whose looks and valor represent the Mexican people—is the one who finally manages to kill him.

8. Robert Wise's film *The Haunting*, based on Shirley Jackson's 1959 novel *The Haunting of Hill House*, shares this lesbian theme. It is interesting that the Jackson's novel shares the same publication date as *Crônica da casa assassinada*.

9. In *Crônica da casa assassinada*, Timóteo's confinement is similar to that of the captive in Brontë's *Jane Eyre*, Rochester's mad first wife, Bertha, who is locked in the attic and later sets fire to Rochester's mansion (also see Mrs. Danvers, who sets fire to Manderley in Du Maurier's *Rebecca*). The landmark study by Sandra Gilbert and Susan Gubar, *The Madwoman in the Attic: The Woman Writer and the Nineteenth-Century Literary Imagination* (1979), is one of the first feminist rereadings of the gothic tradition. The title of the book reflects events in Charlotte Brontë's *Jane Eyre* (1847), which examines the stereotypes of woman as angel or monster.

10. In Rubião's "Petúnia" (1969) we have an example of a vampiric mother and the monstrous feminine combined with imagery reminiscent of Wilde's *The Picture of Dorian Gray* (1890). In this Brazilian story, the evil presence of a mother preserved in a portrait dominates the life of her son Éolo and that of his wife, whom he affectionately calls Petúnia. Before her death, Dona Mineides makes a last request that her portrait be placed above the couple's marital bed. All goes well until the birth of their third daughter, when the paint from the portrait begins to seep down onto the picture frame. However, upon arriving home one afternoon, Éolo finds his daughters strangled and his wife pointing to the portrait, identifying the mother as the perpetrator of the crime. Éolo keeps the murders secret and buries the bodies of his daughters in the garden. Strangely enough, one night he feels compelled to unbury them, at which point they revive and begin to dance. Each morning when he reburies them, flowers sprout on their graves. Every night as his wife sleeps, a rose sprouts from her belly. Éolo plucks it every day before she awakes, but he finally ends up murdering her and burying her in the garden as well. The story closes with Éolo's Sisyphean routine of retouching his mother's portrait, unburying and reburying his daughters, and plucking the profuse flowers growing from his wife's grave, which threaten to alert his neighbors to his crimes. The patriarchal structures embodied by the gothic mansion provokes a response that collapses all categories, animal, botanical, and sexual, which is especially evident in Rubião's story, since life and death, animal and botanical are deeply mixed, indicating the autoimmune response of the vampiric house feeding on humans like plants from the garden.

11. The theme of vampires and same-sex desire became deeply linked in the late 1990s and early 2000s in Brazil. Although Le Fanu's *Carmilla* (1872) portrays the first lesbian vampire, they are actually part of older traditions, as Barbara Creed explores in her study on the monstrous feminine, which she applies to films of female vampirism such as those based on *Carmilla* (60-61) as well as *The Hunger* (1983), which starred Catherine Deneuve and David Bowie (67-72).

12. As a counterexample to female resistance and desire and the return of the parasitical colonial vampire, Carlos Fuentes's short novella *Vlad* comes to mind. It originally appeared in 2004 and returns to the European archetypal predator—not Dracula but the historical figure of Vlad the Impaler (1431–1476). Fuentes creates a dramatic backstory for Vlad, explaining how he becomes an immortal vampire and eventually comes to live in Mexico, where he leaves a trail of bodies, death, and disappearances. As in Agamben's biopolitical model, the victims of Vlad's predations have little hope of prevailing against his power, which dehumanizes them, turning them into "docile bodies"—a resource to be absorbed by global economic and political forces (Campbell 55). For more on the vampiric presence and race in *Vlad*, see Eljaiek-Rodríguez's study (189-209), and for a similar return of the parasitic colonial European vampire in Brazil, see Jacob C. Brown, "Undying (and Undead)" and his analysis of Ivan Jaf's *O vampiro que descobriu o Brasil* (1999) and Nazarethe Fonseca's *Dom Pedro I, vampiro* (2015), in which colonial vampires feed on Afro-descendant women.

13. Among the Brazilian vampire novels are Martha Argel, *Relações de sangue: Vampiros podem estar onde você menos imagina* (2002), *O vampiro de cada um* (2003), and *O vampiro da Mata Atlântica* (2009), as well as the translation of stories with Humberto Moura Neto *O vampiro antes de Drácula* (2008). In addition to her Kaori trilogy, Guilia Moon published a volume of short stories, *Vampiros no espelho* (2004; Vampires in the mirror), and also edited a multi-author collection, *Amor vampiro* (2008; Vampire love). There is an anthology edited by Luiz Roberto Guedes, *O livro vermelho dos vampiros* (2008; The red book of vampires), as well as Nazarethe Fonseca's novel *Kara e Kmam* (2010). Jovane Nunes's *A escrava Isaura e o vampiro* (2010; The slave Isaura and the vampire) is based on the famous abolitionist novel from 1875 by Bernardo Guimarães, *A escrava Isaura* (The slave Isaura).

14. Rob Latham considers vampires to be tools of the commercialization of eternal youth and teenage culture and consumerism in contemporary society in films such as *The Hunger* and *Lost Boys*, while Phil Wegner has looked at television's explorations of issues of gender and sexuality in the *Buffy* and *Angel* television series. Latham's book is called *Consuming Youth: Vampires, Cyborgs, and the Culture of Consumption* (2002) and Wegner's is *Life between Two Deaths, 1989-2001: U.S. Culture in the Long Nineties* (2009).

15. André Vianco's *Os sete* (1999) begins when Tiago and his friends find an oversized silver casket in the waters outside of Porto Alegre, the capital city of

Rio Grande do Sul. Although they find warnings not to open it, they bring it onshore anyway. Later, the divers enlist the help of a friend of theirs, Eliana, an archeology student, to open the casket, but when she inadvertently cuts herself, her blood drips onto one of the bodies, which unbeknownst to them slowly begins to reanimate the vampires one by one. It turns out that they are sixteenth-century Portuguese vampires who were condemned for their crimes and dumped into Brazilian waters in 1506, just six years after the official "discovery" of Brazil in 1500. Having been revived by Eliana's blood, the head vampire, Sétimo, is determined to take her as his mate, and he kidnaps her with the intention of taking her back to Portugal, but the vampires inadvertently create a "hybrid" when Tiago is bitten in an initial skirmish. He becomes an ally of one of the vampires, Miguel, who is determined to kill Sétimo, whose sadism and crimes brought about their entombment. Miguel enlists Tiago to help carry out his revenge on the other vampires, who are eventually destroyed in an explosion.

16. Zárate's novel takes specific details from the original, including characters, but is not overly imitative. The title may also allude to Mexican author Salvador Novo's controversial *La estatua de sal* (The statue of salt) an autobiography/diary about his homosexual exploits that was penned in 1945 but first published in Mexico in 1998 with a prologue by Carlos Monsiváis, who offers a history of gay culture in Mexico. Since Zárate's text was published in 1998, the same year as the publication of the diary, a direct connection is difficult to establish, however, it is tempting to believe that Zárate was aware of Novo and his autobiography.

17. Moon published two sequels in addition to her original *Kaori: Perfume de vampira* (2009): *Kaori 2: Coração de vampira* (2011; Kaori 2: Heart of the vampire) and *Kaori e o samurai sem braço* (2012; Kaori and the one-armed samurai).

18. Moon's Kaori includes many elements of Japanese culture and its own adaptation of the *kyuketsuki* or vampire in popular culture, which is part of the novel's bridging cultures and highlighting immigration within Brazil. For more about Moon, see her interview with Marcello Simão Branco and César Silva in 2006.

BIBLIOGRAPHY

Abreu, Caio Fernando. "A ascensão e queda de Robhéa, manequim e robô." *O ovo apunhalado*. Porto Alegre: Globo, 1975. 31–36.

Agamben, Giorgio. *Homo Sacer: Sovereign Power and Bare Life*. Trans. Daniel Heller Roazen. Stanford, CT: Stanford UP, 1998.

———. "In Praise of Profanation." 2007. *Profanations*. Trans. Jeff Fort. New York: Zone Books, 2015. 73–92.

Alencar, José de. *Iracema*. 1865. Rio de Janeiro: Instituto Internacional do Livro, 1965.

Anderson, Benedict. *Imagined Communities*. 1983. London: Verso, 1991.

Andrade, Carlos Drummond de. "Flor, telefone, moça." *Páginas de sombra: Contos fantásticos brasileiros*. 1951. Ed. Braulio Tavares. Rio de Janeiro: Casa da Palavra, 2003. 20–25.

Andrade, Mário de. *Macunaíma: O heroi sem nenhum caráter*. 1928. São Paulo: Martins, 1978.

Andrade, Oswald de. "Manifesto antropófago." *Vanguardas latino-americanas: Polêmicas, manifestos e textos críticos*. Ed. Jorge Schwartz. São Paulo: Edusp. 140–47.

———. "À marcha das utopias." *Do Pau-Brasil à antropofagia e às utopias*. Rio de Janeiro: Civilização Brasileira, 1978. 145–228.

Arata, Stephen. "The Occidental Tourist: 'Dracula' and the Anxiety of Reverse Colonization." *Victorian Studies* 33.4 (1990): 621–45.

Arboleda-Ríos, Paola. "Homotramas: Actos de resistencia, actos de amor: Construcción de subjetividades LGBT en narraciones del yo latinoamericanas." Diss., U of Florida, 2013.

Arenas, Fernando. *Utopias of Otherness: Nationhood and Subjectivity in Portugal and Brazil*. St. Paul: U of Minnesota P, 2003.

Argel, Martha. *O vampiro da Mata Atlântica*. São Paulo: Ideia Editora, 2009

———. *O vampiro de cada um*. São Paulo: Scortecchi, 2003.

———. *Relações de sangue: Vampiros podem estar onde você menos imagina*. São Paulo: Novo Século, 2002.

Argel, Martha, and Humberto Moura Neto, eds. *O vampiro antes de Drácula*. São Paulo: Aleph, 2008.

Arias, Enrique Desmond. "The Impacts of Differential Armed Dominance of Politics in Rio de Janeiro, Brazil." *Studies in Comparative International Development* 48.3 (Sept. 2013): 263–84.

Arias, Rosario. "Personal Trauma, Romance and Ghostly Traces." *Trauma and Romance in Contemporary British Literature*. Ed. Jean-Michel Ganteau and Susan Onega. New York: Routledge, 2013. 50–67.

Aridjis, Homero. *Ciudad de zombis*. Mexico City: Alfaguara, 2014.

Arreola, Juan José. "Anuncio." *Confabulario total [1941–1961]*. 1961. Mexico City: Fondo de Cultura Económica, 1962. 128–32.

———. "Baby H. P." *Confabulario total [1941–1961]*. 1952. Mexico City: Fondo de Cultura Económica, 1962. 126–27.

Atteberry, Brian. *Decoding Gender in Science Fiction*. New York: Routledge, 2002.

Avelar, Idelber. *The Untimely Present: Postdictatorial Latin American Fiction and the Task of Mourning*. Durham, NC: Duke UP, 1999.

———. "Revisions of Masculinity: Gabeira, Abreu, and Noll." *Literature and Ethics in Contemporary Brazil*. Ed. Vinicius Mariano de Carvalho and Nicola Gavioli. New York: Taylor and Francis, 2017. 183–99.

Avilés, Edgar Omar. "El pantano de los peces esqueleto." *Luna cinema*. Mexico City: Tierra Adentro, 2010. 83–119

Azevedo, Aluísio. *O cortiço*. 1890. São Paulo: Ática, 1990.

Baden, Nancy T. *Muffled Cries: The Writers and Literature in Authoritarian Brazil, 1964–1985*. Lanham, MD: U Press of America, 1999.

Bacigalupi, Paolo. *The Windup Girl*. 2009. San Francisco: Nightshade Books, 2012.

Bailey, Stanley. *Legacies of Race*. Stanford, CA: Stanford UP, 2009.

Barcia, Jacques. "Uma vida possível atrás das barricadas." *Steampunk: Histórias de um passado extraordinário*. Ed. Gianpaolo Celli. São Paulo: Tarja, 2009. 123–40.

Barrera Barrios, Susana Elena. "La mujer fatal en salamandra de Efrén Rebolledo." Undergraduate thesis, Universidad de Sonora, Sonora, Mexico, 2009.

Bartra, Roger. *La jaula de la melancolía: Identidad y metamorfosis del mexicano*. Mexico City: Grijalbo, 1987.

Becerra Acosta, Manuel. "El laboratorio de espíritus." *Los domadores y otras narraciones*. 1945. México, DF: LibroMex, 1960. 111–15.

Beckman, Ericka. *Capital Fictions: The Literature of Latin America's Export Age*. Minneapolis: U of Minnesota P, 2013.

Bef. See Fernández, Bernardo.

Beisecker, Dave. "Nothing but Meat? Philosophical Zombies and Their Cinematic Counterparts." *Race, Oppression and the Zombie: Essays on Cross-Cultural Appropriations of the Caribbean Tradition*. Ed. Christopher M. Moreman and Cory James Rushton. Jefferson, NC: McFarland, 2011. 191–205.

Bell, Andrea L., and Yolanda Molina Gavilán. *Cosmos Latinos: An Anthology of Science Fiction from Latin America and Spain*. Middletown, CT: Wesleyan University Press, 2003.

Bellatín, Mario. *Salón de belleza*. 1994. Mexico City: Tusquets, 2009.

Bermúdez, Maria Elvira. "Soliloquio de un muerto." *Los epígrafes* 10 (1951): 3–7.

Besançon, Guy. "Notas clínicas e psicopatológicas." *Crônica da casa assassinada*. By Lúcio Cardoso. Ed. Mario Carelli. Madrid: UNESCO, 1996. 689–95.

Biehl, João. "Vita: Life in a Zone of Social Abandonment." *Social Text* 19.3 (2001): 131–40.

Boluk, Stephanie, and Wylie Lenz. "Generation Z, the Age of Apocalypse." Introduction. *Generation Zombie: Essays on the Living Dead in Modern Culture*. Ed. Stephanie Boluk and Wylie Lenz. Jefferson, NC: McFarland, 2011. 1–17.

Borges, Dain. "'Puffy, Ugly, Slothful and Inert': Degeneration in Brazilian Social Thought, 1880–1940." *Journal of Latin American Studies* 25.2 (1993): 235–56.

Botting, Fred. "'Monsters of the Imagination': Gothic, Science, Fiction." *A Companion to Science Fiction*. Ed. David Seed. Malden, MA: Blackwell, 2005. 111–26.

Boullosa, Carmen. *Duerme*. Madrid: Santillana, 1994.

———. *Cielos en la tierra*. Mexico City: Alfaguara, 1997.

Braham, Persephone. *From Amazons to Zombies: Monsters in Latin America*. Lewisburg, PA: Bucknell UP, 2015.

———. "The Monstrous Caribbean." *Ashgate Research Companion to Monsters and the Monstrous*. Ed. Asa Simon Mittman and Peter J. Dendle. Burlington, VT: Ashgate, 2016. 17–47.

Branco, Marcello Simão, and César Silva. "Personalidades do ano: Entrevista com Giulia Moon e Martha Argel." *Anuário brasileiro de literatura fantástica: Ficção científica, fantasia e horror no Brasil 2006*. São Bernardo do Campo, Brazil: Edições Hiperespaço, 2007. 57–67.

Brescia, Pablo. "Dos escritores y un bebé: Ansiedades tecnológicas transamericanas" *Revista iberoamericana* 78.238-239 (2012): 91–107.

Brontë, Charlotte. *Jane Eyre*. 1847. New York: Modern Library, 2000.

Brown, J. Andrew. *Cyborgs in Latin America*. New York: Palgrave, 2010.

Brown, Jacob C. "Undying (and Undead) Modern National Myths: Cannibalism and Racial Mixture in Contemporary Brazilian Vampire Fiction." *Alambique* 6.2 (2019): 1–25.

Bukatman, Scott. *Terminal Identity: The Virtual Subject in Postmodern Science Fiction*. Durham, NC: Duke UP, 1993.

Burns, E. Bradford. *A History of Brazil*. New York: Columbia UP, 1980.

Burton, Tara Isabella. "Hadaly, the First Android: Reinstituting the Female Body in Villiers' *Tomorrow's Eve*." *Strange Horizons* 26 August 2013. strangehorizons. com/non-fiction/articles/hadaly-the-first-android-restituting-the-female-body-in-villiers-tomorrows-eve.

Butler, Judith. *Bodies that Matter*. 1993. New York: Routledge, 2011.

——. *Gender Trouble*. 1990. New York: Routledge, 2006.

——. *Precarious Life: The Powers of Mourning and Violence*. London: Verso, 2004.

Calado, Ivanir. "O altar dos nossos corações." *O atlântico tem duas margens: Antologia da novíssima ficção científica portuguesa e brasileira*. Lisboa: Caminho, 1993. 177–206.

Caldeira, Teresa, and James Holston. "Democracy and Violence in Brazil." *Comparative Studies in Society and History* 41.4 (1999): 691–720.

Calogero, Stephen. "Why Positivism Failed in Latin America." *Inter-American Journal of Philosophy* 3.1 (2012): 34–58.

Caminha, Adolfo. *Bom-Crioulo*. 1895. São Paulo: Ática, 1983.

Campbell, Timothy. *Improper Life: Technology and Biopolitics from Heidegger to Agamben*. Minneapolis: U of Minnesota P, 2011.

Campos, Haroldo de. "Da razão antropofágica: A Europa sob a devoração." *Colóquio: Letras*, Vol. 62 (1981): 10–25.

Campos, Victoria E. "Toward a New History: Twentieth-Century Debates in Mexico on Narrating the National Past." *A Twice-Told Tale: Reinventing the Encounter in Iberian/Iberian American Literature and Film*. Ed. Santiago Juan-Navarro and Theodore Robert Young. Newark: U of Delaware P, 2001. 47–64.

Canclini, Nestor García. *Hybrid Cultures: Strategies for Entering and Leaving Modernity*. Trans. Christopher L. Chiappari and Silvia L. López. Minneapolis: U of Minnesota P, 1995

Cano, Luis. *Los espíritus de la ciencia ficción: Espiritismo, periodismo y cultura popular en las novelas de Eduardo Holmberg, Francisco Miralles y Pedro Castera*. Chapel Hill: U of North Carolina P, 2017.

Cardona Peña, Alfredo. "Los hemoglobitas." *Cuentos de magia de misterio y de horror*. Mexico City: Finisterre, 1966. 127–29.

——. "La niña de Cambridge." *Cuentos de magia, de misterio y de terror*. Mexico City: Finisterre, 1966. 115–18.

Cardoso, Lúcio. *Crônica da casa assassinada*. 1959. Critical edition by Mario Carelli. Madrid: UNESCO, 1996.

Carelli, Mario. "A música do sangue." Introduction. *Crônica da casa assassinada*. By Lúcio Cardoso. Ed. Mario Carelli. Madrid: UNESCO, 1996. 723–29.

Carneiro, André. "Brain Transplant." *Cosmos Latinos: An Anthology of Science Fiction from Latin America and Spain*. Ed. Andrea L. Bell and Yolanda Molina Gavilán. Middletown, CT: Wesleyan UP, 2003. 196–200.

——. "Transplante de cérebro. 1978. *A máquina de Hyerónimus e outras histórias*. São Carlos: EDUFSCar, 1997. 115–20.

Caruth, Cathy. *Unclaimed Experience: Trauma, Narrative, and History*. Baltimore, MD: Johns Hopkins UP, 1996.

Carvalho, José Murilo de. "Entrevista." *Quatro autores em busca do Brasil: Entrevistas a José Geraldo Couto*. Rio de Janeiro: Rocco, 2000. 9–29.

Casanova-Vizcaíno, Sandra, and Inés Ordiz, eds. *Latin Amerian Gothic in Literature and Culture*. New York: Routledge, 2017.

Castera, Pedro. *Querens*. 1890. *Impresiones y recuerdos:* Las minas y los mineros; Los maduros dramas; Dramas en un corazón; Querens. Ed. Luis Mario Schneider. Mexico City: Patria, 1987. 387–458.

Castrejón, Eduardo A. (pseud). *Los cuarenta y uno: Novela crítico-social*. 1906. Critical edition by Robert McKee Irwin. Prologue by Carlos Monsiváis. Mexico City: UNAM, 2010.

Castro, Raquel, and Rafael Villegas, eds. *Festín de muertos*. México City: Océano, 2015.

Causo, Roberto de Sousa. *Ficção científica, fantasia e horror no Brasil 1875–1950*. Belo Horizonte, MG: Editora UFMG, 2003.

———. "O novo protótipo." 2008. *Shiroma, matadora ciborgue*. São Paulo: Devir, 2015. 33–44.

———. *O par: Uma novela amazônica*. São Paulo: Humanitas, 2008.

———. "Rosas brancas." 2009. *Shiroma, matadora ciborgue*. São Paulo: Devir, 2015. 13–22.

———. *Shiroma, matadora ciborgue*. São Paulo: Devir, 2015.

———. "Tempestade solar." 2010. *Shiroma, matadora ciborgue*. São Paulo: Devir, 2015, 175–83.

———. ed. *Os melhores contos brasileiros de ficção científica: Fronteiras*. São Paulo: Devir, 2009.

Certeau, Michel de. *The Practice of Everyday Life*. 1984. Trans. Steven Rendall. Berkeley: U of California P, 1988.

Chacek, Karen. *La caída de los pájaros*. México City: Alfaguara, 2014.

———. "La hora de las luciérnagas." *Los viajeros*. Ed. Bef. Mexico City: Planeta, 2010. 161–66.

———. "The Hour of the Fireflies." Trans. Michael J. Deluca. *Three Messages and a Warning*. Ed. Eduardo Jiménez Mayo and Chris N. Brown. Easthampton, MA: Small Beer Press, 2012. 25–30.

Cirne, Roberta. "Boca-de-Ouro." *Sombras do recife: Quadrinhos de terror*. n.d. sombrasdorecife.com.br/quadrinhos/terror-boca-de-ouro.

Clute, John, David Langford, Peter Nicholls, and Graham Sleight. *The Encyclopedia of Science Fiction*. 3rd ed. (online). www.sf-encyclopedia.com.

Coelho Neto, Maximiano Henrique. *Esfinge*. 1908. Porto: Livraria Chardron, 1925.

Cohen, Jeffrey J. "Monster Culture (Seven Theses)." *Monster Theory: Reading Culture*. Minneapolis: U of Minnesota P, 1996. 3–25.

Comaroff, Jean, and John Comaroff. "Alien-Nation: Zombies, Immigrants, and Millennial Capitalism." *South Atlantic Quarterly* 101.4 (2002): 779–805.

Connor, Anne. "Taking the Bite out of the Vagina Dentata: Latin American Women Authors' Fantastic Transformations of the Feline Fatale." *Journal for the Fantastic in the Arts* 24.1 (2013): 41–60.

Cornejo Polar, Antonio. "Mestizaje e hibridez: Los riesgos de las metáforas. Apuntes." *Revista iberoamericana* 180 (1997): 341–44.

Corral Rodríguez, Fortino, and Nubia Uriarte Montoya. "Elementos para una aproximación simbólica a 'El huésped' de Amparo Dávila." *ConNotas: Revista de crítica y teoría literaria* 6.11 (2008): 211–21.

Craft, Christopher. "'Kiss Me with Those Red Lips': Gender and Inversion in Bram Stoker's *Dracula*." *Dracula*. By Bram Stoker. 1897. Ed. Nina Auerbach and David J. Skal. New York: Norton, 1997. 444–59.

Creed, Barbara. *The Monstrous Feminine: Film, Feminism and Psychoanalysis*. London: Routledge, 1993.

Cruls, Gastão. *A Amazonia misteriosa*. 1925. São Paulo: Saraiva, 1958.

Crump, Marty. *Eye of Newt and Toe of Frog, Adder's Fork and Lizard's Leg: The Lore and Mythology of Amphibians and Reptiles*. Chicago: U of Chicago P, 2015.

Cuenca, J. P. *O único final feliz para uma história de amor é um acidente*. São Paulo: Companhia das Letras, 2010.

Cuevas, Alejandro. "El vampiro." *Cuentos macabros*. Mexico City: Garrido y Hermanos, 1911. 67–80.

Dalton, David. "Antropofagia, Calibanism, and the Post-Romero Zombie: Cannibal Resistance in Latin America and the Caribbean. *Alambique: Revista académica de ciencia ficción y fantasia / Jornal acadêmico de ficção científica e fantasia* 6.1 (2018): 1–21.

———. *Mestizo Modernity: Race, Technology and the Body in Postrevolutionary Mexico*. Gainesville: U of Florida P, 2018.

DaMatta, Roberto. "For an Anthropology of the Brazilian Tradition, Ou, A virtude está no meio." *The Brazilian Puzzle: Culture on the Borderlands of the Western World*. Ed. David J. Hess and Roberto DaMatta. New York: Columbia UP, 1995. 270–92.

———. *Carnavais, malandros e heróis: Para uma sociologia do dilema brasileiro*. 1979. Rio de Janeiro: Guanabara, 1990.

———. *O que faz o Brasil, Brasil?* Rio de Janeiro: Rocco, 1984.

Damián Miravete, Gabriela. "La mano izquierda de la ciencia ficción mexicana." *Letras libres* 1 Oct. 2018. www.letraslibres.com/mexico/revista/la-mano-izquierda-la-ciencia-ficcion-mexicana.

Dávila, Amparo. "El huésped." *Cuentos reunidos*. 1959. Mexico City: Fondo de Cultura Económica, 2009. 19–23.

De Castro, Juan E. *Mestizo Nations: Culture, Race and Conformity in Latin American Literature*. Tucson: U of Arizona P, 2002.

De Cicco, Mark. "'More than Human': The Queer Occult Explorer in the Fin-de-Siècle." *Journal of the Fantastic in the Arts* 23.1 (2012): 4–24.

Del Picchia, Paulo Menotti. *A filha do Inca*. 1930. São Paulo: Saraiva, 1949.

———. *Cummunká*. 1938. São Paulo: Martins, 1958.

———. *Kalum*. 1936. São Paulo: Saraiva, n/d.

Derrida, Jacques. "The Law of Genre." Trans. Avital Ronell. *On Narrative*. Ed. W. J. T. Mitchell. Chicago: U of Chicago P, 1980. 51–77.

———. "Structure, Sign and Play in the Discourse of the Human Sciences." *The Structuralist Controversy: The Languages of Criticism and the Sciences of Man*. 1970. Ed. Richard Macksey and Eugenio Donato. Baltimore: Johns Hopkins UP, 1979. 247–64.

Du Maurier, Daphne, *Rebecca*. 1938. New York: Avon, 1971.

Dunbar, David L. "Unique Motifs in Brazilian Science Fiction." Diss., University of Arizona, 1976.

Echeverría, Bolívar. *La modernidad de lo barroco*. Mexico City: Era, 1998.

———. "El ethos barroco." *Modernidad, mestizaje cultura, ethos barroco*. Mexico City: UNAM, 1994. 13–36.

Edmonds-Poli, Emily, and David Shirk. *Contemporary Mexico Politics*. 2nd ed. Lanham, MD: Rowman, Littlefield, 2012.

Edwards, Erika. "Slavery in Argentina." *Oxford Bibliographies*, 27 June 2017. DOI: 10.1093/OBO/9780199766581–0157.

Edwards, Justin D., and Sandra Vasconcelos, eds. *Tropical Gothic in Literature and Culture: The Americas*. New York: Routledge, 2016.

Eljaiek-Rodríguez, Gabriel Andrés. "Selva de fantasmas: Tropicalización de lo gótico en la literatura y el cine latinoamericanos." Diss., Emory University, 2012.

Ellison, Fred P. *Alfonso Reyes e o Brasil: Um mexicano entre cariocas*. Trans. Cristine Ajuz. Rio de Janeiro: Topbooks, 2002.

Esquinca, Bernardo. "La otra noche de Tlatelolco." *Festín de muertos*. Ed. Raquel Castro and Rafael Villegas. Mexico City: Oceano, 2015. 17–26.

Esposito, Roberto. *Bíos: Biopolitics and Philosophy*. Trans. Timothy Campbell. University of Minnesota Press, 2008.

———. *Communitas: The Origin and Destiny of Community*. Trans. Timothy Campbell. Stanford, CA: Stanford UP, 2010.

———. *Immunitas: The Protection and Negation of Life*. Trans. Zakiya Hanafi. Malden, MA: Polity, 2011.

Evans, Peter, and Gary Gereffi. "Foreign Investment and Dependent Development: Comparing Brazil and Mexico." *Brazil and Mexico: Patterns in Late Development*. Ed. Sylvia Ann Hewlett and Richard S. Weinert. Philadelphia, PA: Institute for Study of Human Issues, 1982. 111–68.

Fausto, Boris. *A história concisa do Brasil*. São Paulo: Edusp, 2001.

Fernández Delgado, Miguel Ángel. "Introducción." *Visiones periféricas: Antología de la ciencia ficción mexicana*. Mexico City: Lumen, 2000. 7–17.

———. "Mexico." *The Encyclopedia of Science Fiction*, 2 Feb. 2017. Ed. John Clute, David Langford, Peter Nicholls, and Graham Sleight Gollancz. www.sf-encyclopedia.com/entry/mexico.

Fernández, Bernardo [Bef], ed. *Los viajeros: 25 años de ciencia ficción mexicana*. Mexico City: SM Ediciones, 2010.

———. *Gel azul*. Mexico City: Santillana, 2009.

———. ed. *La imaginación: La loca de la casa*. Mexico City: Conaculta, 2015.

———. "Las últimas horas de los últimos días." 2006. *Los viajeros: 25 años de ciencia ficción mexicana*. Ed. Bernardo Fernández [Bef]. Mexico City: SM Ediciones, 2010. 168–79.

Fernández, Bernardo, and Gerardo Sifuentes. "(e)." *Visiones periféricas: Antología de ciencia ficción mexicana*. Ed. Miguel Ángel Fernández Delgado. Mexico City: Lumen, 2001. 173–86.

Fernandez, Cid. "Julgamentos." *Tríplice universo*. Ed. Roberto de Sousa Causo. São Paulo: GRD, 1993. 70–152.

Ficker, Sandra Kuntz. "Institutional Change and Foreign Trade in Mexico, 1870–1911." *The Mexican Economy: Essays on the Economic History of Institutions, Revolution, and Growth*. Ed. Jeffery L. Bortz and Stephen Haber. Stanford, CA: Stanford UP, 2002. 161–204.

Fischer, Brodwyn. *A Poverty of Rights: Citizenship and Inequality in Twentieth Century Rio de Janeiro*. Stanford, CA: Stanford UP, 2008.

Fischer-Hornung, Dorothea, and Monika Mueller. *Vampires and Zombies: Transcultural Migrations and Transnational Interpretations*. Jackson: UP of Mississippi, 2016.

Flammarion, Camille. *Lumen*. Trans. and ed. Brian Stableford. Middletown, CT: Wesleyan UP, 2002.

Flaubert, Gustave. "Quidquid volueris." *Passion et vertu; Quidquid volueris; Mémories d'un fou; Un parfum à sentir*. Paris: Safrat, 1990. 37–68.

Fonseca, Nazarethe. *Dom Pedro I, vampiro*. Mexico City: Planeta, 2015.

———. *Kara e Kmam*. São Paulo: Aleph, 2010.

Fonseca, Rubem. *A grande arte*. 1983. São Paulo: Companhia das Letras, 1990.

Foucault, Michel. *Discipline and Punish: The Birth of the Prison*. Trans. Alan Sheridan. New York: Vintage, 1978.

———. *The History of Sexuality*. Trans. Robert Hurley. New York: Vintage, 1980.

———. *Society Must Be Defended: Lectures at the Collège de France 1975–1976*. Trans. David Macey. New York: Picador, 2003.

Franco, Jean. "La Malinche: From Gift to Sexual Contract." *Critical Passions: Selected Essays*. Ed. Mary Louise Pratt and Kathleen Newman. Durham, NC: Duke University Press, 1999. 66–82.

French, William E., and Katherine Elaine Bliss. "Gender, Sexuality, and Power since Independence." Introduction. *Gender Sexuality, and Power in Latin America since Independence*. Ed. William E. French and Katherine Elaine Bliss. Lanham, MD: Rowman and Littlefield, 2007. 1–30.

Freud, Sigmund. "The Uncanny." *Standard Edition of the Complete Psychological Works*. Trans. James Strachey. London: Hogarth Press, 1955. 217–55.

Freyre, Gilberto. *Casa-grande e senzala: Formação da família brasileira sob o regime de economia patriarcal*. 2 vols. 1933. Rio Janeiro: José Olympio, 1952.

———. "Boca-de-ouro." 1955. *Assombrações do Recife Velho: Algumas notas históricas e outras tantas folclóricas em torno do sobrenatural no passado recifense.* Introd. Newton Moreno. São Paulo: Global, 2012. 62–63.

Fuentes, Carlos. "Tlactocatzine en el jardín de Flandes." *Los días enmascarados.* 1954. Mexico City: Era, 1982. 34–45.

———. *Vlad.* 2004. Mexico City: Alfaguara, 2012.

Gamio, Miguel. *Forjando patria.* 1916. Mexico City: Porrúa, 1992.

Gandler, Stefan. *Critical Marxism in Mexico: Adolfo Sánchez Vásquez and Bolívar Echeverría.* Trans. George Ciccariello-Maher and Stefan Gandler. Leiden: Brill, 2014.

García, Hernán Manuel. "Carne eres y en máquina te convertirás: El cuerpo post-humano en la primera calle de la soledad de Gerardo Porcayo." *Polifonía* 4.1 (2014): 1–14.

García Gutiérrez, Georgina. "Tlactocatzine en el jardín de Flandes." *Los disfraces: La obra mestiza de Carlos Fuentes.* Mexico City: Colegio de Mexico, 2000. 43–59.

Gaspari, Elio. *A ditadura escancarada.* São Paulo, Companhia das Letras, 2002.

Gibson, William. *Neuromancer.* New York: Ace, 1984.

Gilbert, Sandra, and Susan Gubar. *The Madwoman in the Attic: The Woman Writer and the Nineteenth-Century Literary Imagination.* 1979. New Haven, CT: Yale UP, 2000.

Ginway, M. Elizabeth. "The Amazon in Brazilian Speculative Fiction: Utopia and Trauma." *Alambique: Revista académica de ciencia ficción y fantasia / Jornal acadêmico de ficção científica e fantasia* 3.1 (2015): 1–15.

———. "The Body in Brazilian Science Fiction: Implants and Cyborgs." *New Boundaries in Political Science Fiction.* Ed. Donald M. Hassler and Clyde Wilcox. Columbia: U of South Carolina P, 2008. 198–211.

———. *Brazilian Science Fiction: Cultural Myths and Nationhood in the Land of the Future.* Lewisburg: PA: Bucknell UP, 2004.

———. "Eating the Past: Proto-Zombies in Brazilian Fiction 1900–1955." *Alambique: Revista académica de ciencia ficción y fantasia / Jornal acadêmico de ficção científica e fantasia* 6.1 (2018): 1–14.

———. "Eugenia, a mulher e a política na literatura utópica brasileira (1909–1929)." *Cartografias para a ficção científica mundial: Cinema e literatura.* Ed. Alfredo Suppia. São Paulo: Alameda, 2015. 209–22.

———. "Machado's Tales of the Fantastic: Allegory and the Macabre." *Emerging Dialogues on Machado de Assis.* Ed. Lamonte Aidoo and Daniel F. Silva. New York: Palgrave, 2016. 211–22.

———. "Posthumans in Brazilian Cyberpunk and Steampunk." *Latin American Speculative Fiction.* Ed. Debra Ann Castillo and Liliana Colanzi. Spec. issue of *Paradoxa,* no. 30 (2018): 233–50.

———. "*O presidente negro* [The black president]: Eugenics, Race and Gender in Brazil's Corporate State." *Black and Brown Planets: The Politics of Race in Science Fiction.* Ed. Isiah Lavender III. Jackson: U of Mississippi P, 2014. 131–45.

———. "Recent Brazilian Science Fiction and Fantasy Written by Women." *Foundation* 26.99 (2007): 49–62.

———. "Simios, ciborgues e reptiles: La oviparidad en obras de escritoras latinoamericanas de ciencia ficción y fantasia." *Revista iberoamericana* 33.259-260 (2017): 645–56.

———. "Teaching Latin American Science Fiction and Fantasy in English: A Case Study." *Teaching Science Fiction*. Ed. Andy Sawyer and Peter Wright. New York: Palgrave, 2011. 179–201.

———. "Transgendering in Brazilian Speculative Fiction from Machado de Assis to the Present." *Luso-Brazilian Review* 47.1 (2010): 40–60.

———. "Vampires, Werewolves, and Strong Women: Alternate Histories or the Re-writing of Race and Gender in Brazilian History." *Extrapolation* 44.3 (2003): 283–95.

Ginway, M. Elizabeth, and Roberto de Sousa Causo. "Discovering and Rediscovering Brazilian Science Fiction: An Overview." *Extrapolation* 51.1 (2010): 13–39.

Girard, René. "The Plague in Literature and Myth." *To Double Business Bound: Essays on Literature, Mimesis and Anthropology*. Baltimore, MD: Johns Hopkins UP, 1978. 136–54.

———. *Violence and the Sacred*. Trans. Patrick Gregory. Baltimore: Johns Hopkins UP, 1977.

Gledson, John. "*Quincas Borba*." *Machado de Assis: Ficção e história*. Trans. Sônia Coutinho. Rio de Janeiro: Paz e Terra, 1986. 58–113.

González, Emiliano. "La herencia de Cthulhu." *Los sueños de la bella durmiente*. Mexico City: Joaquín Mortiz, 1978. 62–73.

———. "Rudisbroeck o los autómatas." *Los sueños de la bella durmiente*. Mexico City: Joaquín Mortiz, 1978. 15–54.

González, Jennifer. "Envisioning Cyborg Bodies: Notes from Current Research." *The Gendered Cyborg: A Reader*. Ed. Gill Kirkup, Linda James, Kathryn Woodward, and Fiona Hovenden. London: Routledge, 2000. 58–73.

Gordon, Joan. "The Responsibilities of Kinship: The Amborg Gaze in Speculative Fictions about Apes." *Extrapolation* 57.3 (2016): 251–64.

Gordus, Andrew M. "The Vampiric and the Urban Space in Dalton Trevisan's *O vampiro de Curitiba*." *Rocky Mountain Review* 52.1 (1998): 13–26.

Graham, Douglas H. "Mexican and Brazilian Economic Development: Legacies, Patterns, and Performance." *Brazil and Mexico: Patterns in Late Development*. Ed. Sylvia Ann Hewlett and Richard S. Weinert. Philadelphia, PA: Institute for Study of Human Issues, 1982. 13–55.

Green, James N. "Doctoring the National Body: Gender, Race and Eugenics, and the Invert in Urban Brazil, ca. 1920–1945. *Gender Sexuality, and Power in Latin America since Independence*. Ed. William E. French and Katherine Elaine Bliss. Lanham, MD: Rowman and Littlefield, 2007. 187–211.

———. *Beyond Carnival: Male Homosexuality in Twentieth-Century Brazil*. Chicago: University of Chicago Press, 1999

Grosz, Elizabeth. *Volatile Bodies: Toward a Corporeal Feminism.* Bloomington: Indiana UP, 1994.

Guedes, Luiz Roberto, ed. *O livro vermelho dos vampiros.* São Paulo: Devir, 2008.

Guerrero Heredia, Marco Vladimir. "El narcogótico mexicano." Diss., Pontificia Universidad Católica de Chile, 2018.

Gutiérrez Mouat, Ricardo. "Gothic Fuentes." *Revista hispánica moderna* 1–2 (2004): 297–313.

Gutiérrez, Gerardo. "Identity Erasure and Demographic Impacts of the Spanish Caste System on the Indigenous Populations of Mexico." *Beyond Germs: Native Depopulation in North America.* Ed. Catherine M. Cameron, Paul Kelton, and Alan C. Swedlund. Tucson: U of Arizona P, 2015. 119–45.

Halberstam, Judith. *The Queer Art of Failure.* Durham, NC: Duke UP, 2011.

———. *Skin Shows: Gothic Horror and the Technology of Monsters.* Durham, NC: Duke UP, 1995.

Haraway, Donna. "The Cyborg Manifesto." *Simians, Cyborgs and Women: The Reinvention of Nature.* 1985. London: Free Association Books, 1991. 149–181.

———. "The Biopolitics of Postmodern Bodies: Constitutions of Self in Immune System Discourse." *Biopolitics: A Reader.* Ed. Timothy Campbell and Adam Sitze. Durham, NC: Duke UP, 2013. 274–309.

Harpold, Terry. "The End Begins: John Wyndham's Zombie Cozy." *Generation Zombie: Essays on the Living Dead in Modern Culture.* Ed. Stephanie Boluk and Wylie Lenz. Jefferson, NC: McFarland, 2011. 156–64.

Hayles, N. Katherine. *How We Become Posthuman: Virtual Bodies in Cybernetics, Literature, and Informatics.* Chicago: U of Chicago P, 1999.

Haywood Ferreira, Rachel. *The Emergence of Latin American Science Fiction.* Middletown, CT: Wesleyan UP, 2011.

Hershfield Joanne. "Domestic Technologies: Gender, Technology and Mexican Housewives 1930–1950." *Technology and Culture in Twentieth-Century Mexico.* Ed. Araceli Tinajero and J. Brian Freeman. Tuscaloosa: University of Alabama Press, 2013. 55–69.

Hess, David J. *Spirits and Scientists: Ideology, Spiritism and Brazilian Culture.* University Park: Pennsylvania State UP, 1991.

Hewlett, Sylvia Ann, and Richard S. Weinert. "The Characteristics and Consequences of Late Development in Brazil and Mexico." *Brazil and Mexico: Patterns in Late Development.* Ed. Sylvia Ann Hewlett and Richard S. Weinert. Philadelphia, PA: Institute for Study of Human Issues, 1982. 1–11.

Hind, Emily. *Femenism and the Mexican Woman Intellectual from Sor Juana to Poniatowska: Boob Lit.* New York: Palgrave McMillan, 2010.

Hoffmann, E. T. A. "The Sandman." 1816. *Horror Stories: Classic Tales from Hoffmann to Hodgson.* Ed. Darryl Jones. New York: Oxford UP, 2014. 3–33.

Hoeg, Jerry. *Science, Technology and Latin American Narrative in the Twentieth Century.* Bethlehem, PA: Lehigh UP, 2000.

Holanda, Sérgio Buarque de. *Raízes do Brasil.* 1936. Rio de Janeiro: José Olympio, 1982.

Holston, James. "Insurgent Citizenship in an Era of New Global Peripheries." *City and Society* 21.2 (2009): 245–67.

Honores, Elton. *Los que moran en las sombras: Asedios al vampiro en la literatura peruana.* Lima: El Lamparero Alucinado, 2010.

Hook, Derek William. "Lacan, the Meaning of the Phallus and the Sexed Subject." *The Gender of Psychology.* Ed. Tamara Shefer, Floretta Boonzaier, and Peace Kiguwa. Cape Town: Juta Academic Publishing, 2006. 60–84.

Irwin, Robert McKee. "Lo que comparte el positivismo con el modernismo mexicano: El hemafroditismo, la bestialidad y la necrofilia." *XIX Signos Literarios* 2.4 (2006): 63–80.

———. *Mexican Masculinities.* Minneapolis, MN: U of Minnesota P, 2003.

Islam, Gazi. "Can the Subaltern Eat? Anthropophagic Culture as a Lens on Post-Colonial Theory." *Organization* 19.2 (2011): 160–80.

Jackson, Rosemary. *Fantasy: The Literature of Subversion.* London: Methuen, 1981.

Jackson, Shirley. *The Haunting of Hill House.* 1959. Introd. Guillermo del Toro. London: Penguin Horror, 2014.

Jaf, Ivan. *O vampiro que descobriu o Brasil.* São Paulo: Ática, 1999.

Jameson, Fredric. "Postmodernism or the Cultural Logic of Capitalism." *New Left Review* 146 (1984): 53–92.

Janzen, Rebecca. *The National Body in Mexican Literature.* New York: Palgrave Macmillan, 2015.

Jáuregui, Carlos. *Canibalismo, calibanismo, antropofagia cultural y consumo en América Latina.* La Habana: Casa de las Américas, 2005.

Jones, Julie. "Paulo Barreto's 'O bebê de tarlatana rosa': A Carnival Adventure." *Luso-Brazilian Review* 24.1 (1987): 27–33.

Jorge, Miguel. "Véspera de pânico." *Avarmas.* São Paulo: Ática, 1980. 28–47.

Jrade, Cathy. *Delmira Agustini, Sexual Seduction, and Vampiric Conquest.* New Haven, CT: Yale UP, 2012.

Kafka, Franz. *The Metamorphosis.* 1915. Trans. Stanley Corngold. New York: Bantam, 2004.

Kamen, Deborah. "Naturalized Desires in the Metamorphosis of Iphis." *Helios* 39.1 (2012): 21–36.

King, Edward, and Joanna Page. *Posthumanism and the Graphic Novel in Latin America.* London: University College of London, 2017.

Kittleson, Roger. *The Country of Football: Soccer and the Making of Modern Brazil.* Berkeley: U of California P, 2014.

Klor de Alva, J. Jorge. "The Postcolonial Latin American Experience: A Reconsideration of 'Colonialism,' 'Post-Colonialism' and 'Mestizaje.'" *After Colonialism, Imperial Histories and Postcolonial Displacements.* Ed. Gyan Prakash. Princeton, NJ: Princeton UP, 1995. 241–75.

Kordas, Ann. "New South, New Immigrants, New Women, New Zombies: The Historical Development of the Zombie in American Culture." *Race, Oppres-*

sion and the Zombie: Essays on Cross-Cultural Appropriations of the Caribbean Tradition. Ed. Christopher M. Moreman and Cory James Rushton. Jefferson, NC: McFarland, 2011. 15–30.

Krause, James R. "Enucleated Eyes in 'Sem olhos' and 'O capitão Mendonça' by Machado de Assis." *Border Crossings: Boundaries of Cultural Interpretation.* Ed. Pablo Martínez Diente and David Wiseman. Nashville, TN: Vanderbilt UP, 2009. 51–63.

———. "Undermining Authoritarianism: Retrofitting the Zombie in 'Seminário dos ratos' by Lygia Fagundes Telles." *Alambique* 6.1 (2018): 1–22.

Kristeva, Julia. *Powers of Horror: An Essay on Abjection.* Trans. Léon S. Roudiez. New York: Columbia UP, 1982.

Kuhnheim, Jill. "Postmodern Feminist Nomadism in Carmen Boullosa's *Duerme.*" *Letras Femeninas* 27.2 (2001): 8–23.

Lampert-Weissig, Lisa. "Taking Dracula's Pulse: Historicizing the Vampire." *The Vampire Goes to College.* Ed. Lisa A. Nevárez. Jefferson, NC: McFarland, 2014. 32–43.

Lara Pardo, Luis. "Prólogo." *Los domadores y otras narraciones.* By Manuel Becerra Acosta. 1945. Mexico City: Libro Mex, 1960. 11–14.

Larbalestier, Justine. *The Battle of the Sexes in Science Fiction.* Middletown, CT: Wesleyan UP, 2002.

Larson, Ross. *Fantasy and Imagination in the Mexican Narrative.* Tempe: Arizona State UP, 1977.

Latham, Rob. *Consuming Youth: Vampires, Cyborgs, and the Culture of Consumption.* Chicago: U of Chicago P, 2002.

———. "Sextrapolation in New Wave Science Fiction." *Queer Universes: Sexualities in Science Fiction.* Ed. Wendy Gay Pearson, Veronica Hollinger, and Joan Gordon. Liverpool UP, 2010. 52–71.

Lauro, Sarah Juliet, and Karen Embry. "A Zombie Manifesto: The Nonhuman Condition in the Age of Advanced Capitalism." *Boundary 2* (Spring 2008): 85–108.

Lear, John. *Picturing the Proletariat: Artists and Labor in Revolutionary Mexico, 1908–1940.* Austin: University of Texas Press, 2017.

Le Fanu, Joseph Sheridan. *Carmilla.* 1872. Createspace, 2013.

Le Guin Ursula K. *The Left Hand of Darkness.* 1969. New York: Ace Books, 2000.

Lehnen, Leila. *Citizen and Crisis in Contemporary Brazilian Literature.* New York: Palgrave Macmillan, 2013.

Lewis, Colin M. "States and Markets in Latin America: The Political Economy of Economic Interventionism." Conference Presentation. GEHN Conference, Sept. 2003. Working Paper No. 09/05, Jan. 2005. eprints.lse.ac.uk/22483/1/wp09.pdf.

Leones, André de. *Dentes negros.* Rio de Janeiro, Rocco, 2011.

Lima Barreto, Afonso Henrique de. *Diário do hospício e o cemetério dos vivos.* 1953. São Paulo: Companhia das Letras, 2017

———. "A nova Califórnia." 1911. *Os melhores contos brasileiros de ficção científica: Fronteiras*. Ed. Roberto de Sousa Causo. São Paulo: Devir, 2009. 25–34.

———. *Triste fim de Policarpo Quaresma*. 1911. Edição crítica. Paris: UNESCO, 1997.

Lobato, José Bento Monteiro. "Café, café." *Cidades mortas*. 1919. São Paulo: Brasiliense, 1959. 177–82.

Lockhart, Darrell B., ed. *Latin American Science Fiction: An A-to-Z Guide*. Westport, CT: Greenwood P, 2004.

Lodi-Ribeiro, Gerson. "Assessor para assuntos fúnebres." 1999. *Outros Brasis*. São Paulo: Mercuryo, 2006. 93–125.

———. *Aventuras de um vampiro de Palmares*. São Paulo: Draco, 2014.

———. "O capitão Diabo das Geraes." *Aventuras de um vampiro de Palmares*. São Paulo: Draco, 2014. 130–68.

———. "Longa viagem para casa." *A viagem*. Ed. Silvana Moreira and António de Macedo. Cascais: Simetria, 2000. 81–97.

———. "Os anos em que fui Carla." Prefácio. *Histórias de ficção científica por Carla Cristina Pereira*. São Paulo: Draco, 2012. 8–17.

———. "O vampiro de Nova Holanda." *O vampiro de Nova Holanda*. Lisboa: Caminho, 1998. 65–116.

———. "Xochiquetzal, ou, A esquadra da vingança." *Histórias de ficção científica por Carla Cristina Pereira*. 2000. São Paulo: Draco, 2012. 64–84.

———. "Xochiquetzal." Trans. David Alan Prescott. Spec. issue of *Altair* (Australia) 6 & 7 (2000): 70–81.

Lopes, Denilson. "Silviano Santiago, estudos culturais e estudos LBGTS no Brasil." *Revista iberoamericana* 74.225 (2008): 943–57.

López Castro, Ramón. *Expedición a la ciencia ficción mexicana*. Mexico City: Lectorum, 2001.

López-Lozano, Miguel. *Utopian Dreams, Apocalyptic Nightmares: Globalization in Recent Mexican and Chicano Narrative*. West Lafayette, IN: Purdue UP, 2008.

Love, Joseph. *Crafting the Third World: Theorizing Underdevelopment in Rumania and Brazil*. Stanford, CA: Stanford UP, 1996.

Loudis, Jessica. "El Chapo: What the Rise and Fall of the Kingpin Reveals about the War on Drugs." *Guardian*, 7 June, 2019. www.theguardian.com/world/2019/jun/07/el-chapo-the-last-of-the-cartel-kingpins.

Luckhurst, Roger. "Cultural Governance, New Labour, and the British SF Boom." *Science Fiction Studies* 30.3 (2003): 417–35.

Lund, Joshua. *The Mestizo State: Reading Race in Modern Mexico*. Minneapolis: U of Minnesota P, 2012.

Machado de Assis, Joaquim Maria. 1884. "As academias de Sião." *Histórias sem data. Obra completa: Conto e teatro*, vol. 2. Rio de Janeiro: Nova Aguilar, 1986. 468–73.

———. "O capitão Mendonça." 1870. *Contos fantásticos de Machado de Assis*. Ed. Raimundo Magalhães Júnior. Rio de Janeiro: Bloch, 1973. 181–203.

———. *Memórias Póstumas de Bras Cubas*. 1881. São Paulo: Ática, 1981.

———. *The Posthumous Memoirs of Brás Cubas*. Trans. Flora Thomson-DeVeaux. New York: Penguin Classics, 2020.

———. *Quincas Borba*. 1890. São Paulo: Ática, 1982.

———. "A vida eterna." 1870. *Contos fantásticos de Machado de Assis*. Ed. Raimundo Magalhães Júnior. Rio de Janeiro: Bloch, 1973. 99–112.

Macías-González, Víctor M. "The Transnational Homophile Movement and the Development of Domesticity in Mexico City's Homosexual Community 1930–70." *Gender, Imperialism and Global Exchanges*. Ed. Stephanie S. Miescher, Michele Mitchell, and Naoko Shibusawa. Oxford, UK: Wiley Blackwell, 2015. 132–57.

Mahoney, Phillip. "Mass Psychology and the Analysis of the Zombie: From Suggestion to Contagion." *Generation Zombie: Essays on the Living Dead in Modern Culture*. Ed. Stephanie Boluk and Wylie Lenz. Jefferson, NC: McFarland, 2011. 113–29.

Martínez Morales, José Luis. "¿Vampiros mexicanos o vampiros en México?" *Lejana: Revista crítica de narrativa breve* 4 (2012): 1–10.

Matangrano, Bruno Anselmi, and Enéias Tavares. *Fantástico brasileiro: O insólito literário do romantismo ao fantasismo*. Curitiba: Arte e Letra, 2018.

Mbembe, J. A. "Necropolitics." Trans. Libby Meintjes. *Public Culture* 15.1 (2003): 11–40.

McCann, Bryan. *Hello, Hello Brazil: Popular Music and the Making of Modern Brazil*. Durham, NC: Duke UP, 2004.

McLuhan, Marshall. *Understanding Media: The Extensions of Man*. New York: McGraw Hill, 1964.

Melo, Alfredo César. "Saudosismo e crítica social em *Casa grande & senzala*: A articulação de uma política da memória e de uma utopia." *Estudos avançados* 23.67 (2009): 279–96.

Meireles da Silva, Alexander. "Um monstro entre nós: A ascensão da literatura gótica do Brasil da belle époque." *A revista do SELL* 2.1 (2010): 1–21.

Merrim, Stephanie. "Catalina de Erauso: From Anomaly to Icon." *Coded Encounters: Writing, Gender, and Ethnicity in Colonial Latin America*. Ed. Francisco Cevallos et al. Amherst: U. of Massachusetts P, 1994. 177–205.

Messinger Cypess, Sandra. "'Mother' Malinche and Allegories of Gender, Ethnicity and National Identity in Mexico." *Feminism, Nation and Myth: La Malinche*. Ed. Ronaldo Romero and Amanda Nolacea Harris. Houston: Arte Público, 2005. 14–27.

Meter, Alejandro. "Argentina in the Era of Mass Immigration." *Oxford Bibliographies*, last modified 30 June 2014. DOI: 10.1093/OBO/9780199766581–0163.

Miliotes, Diana. *José Guadalupe Posada and the Mexican Broadside*. Chicago: Art Institute of Chicago, 2006.

Milner, Andrew. *Locating Science Fiction*. Liverpool, UK: Liverpool UP, 2012.

Molina-Gavilán, Yolanda, Andrea L. Bell, Miguel Ángel Fernández-Delgado, M. Elizabeth Ginway, Luis Pestarini, and Juan Carlos Toledano Redondo.

"Chronology of Latin American Science Fiction, 1775–2005." *Science Fiction Studies* 34.3 (2007): 369–431.

Monteiro, Luciana C. "Mulheres, mães e alienígenas: Subjetividade feminina na ficção científica de Dinah Silveira de Queiroz." *Hispania* 96.4 (2013): 724–34.

Moon, Giulia, ed. *Amor vampiro*. São Paulo: Giz editorial, 2008.

———. *Kaori 2: Coração de vampira*. São Paulo: Giz editorial, 2011.

———. *Kaori: Perfume de vampira*. São Paulo: Giz Editorial, 2009.

———. *Kaori e o samurai sem braço*. São Paulo: Giz Editorial, 2012.

———. *Vampiros no espelho*. São Paulo: Landy, 2004.

Moraña, Mabel. *El monstruo como máquina de guerra*. Madrid: Iberoamericana-Vervuert, 2017.

Moreira, Paulo. *Literary and Cultural Relations between Brazil and Mexico: Deep Undercurrents*. New York: Palgrave Macmillan, 2013.

Moreira-Almeida, Alexander, Angélica Silva de Almeida, and Francisco Lotufo Neto. "Spiritist Madness in Brazil." *History of Psychiatry* 16.1 (2005): 5–25.

Moreland, Sean. "Shambling towards Mount Improbable to Be Born: American Evolutionary Anxiety and the Hopeful Monsters of Matheson's *I Am Legend* and Romero's *Dead Films*." *Generation Zombie: Essays on the Living Dead in Modern Culture*. Ed. Stephanie Boluk and Wylie Lenz. Jefferson, NC: McFarland, 2011. 77–89.

Moreman, Christopher M., and Cory James Rushton. "Race, Colonialism and the Evolution of the 'Zombie.'" Introduction. *Race, Oppression and the Zombie: Essays on Cross-Cultural Appropriations of the Caribbean Tradition*. Jefferson, NC: McFarland, 2011. 1–14.

Moretti, Franco. "The Dialectic of Fear." *New Left Review* 136 (Nov-Dec. 1982): 67–82.

Moussong, Lazlo. *Castillos en la letra*. Xalapa: Universidad Veracruzana, 1986.

Muñoz Zapata, Juan Ignacio. "Narrative and Dystopian Forms of Life in Mexican Cyberpunk Novel *La primera calle de la soledad*." *Science Fiction, Imperialism and the Third World: Essays on Postcolonial Literature and Film*. Ed. Ericka Hoagland and Reema Sarwal. Jefferson, NC: McFarland, 2010. 188–201.

Nascimento, Abdias do. "The Myth of Racial Democracy." *Brazil Reader*. Ed. Robert M. Levine and John J. Crocitti. Durham, NC: Duke UP, 1999. 379–81.

Nemi Neto, João. "The Anthropophagic Queer: A Study on Abjected Bodies and Brazilian Queer Theory." Diss., City University of New York, 2015.

Nervo, Amado. "El donador de almas." 1899. *La revista quincenal: Revista literaria* 3.5 (1920): 3–79.

Nevárez, Lisa A., ed. *The Vampire Goes to College*. Jefferson, NC: McFarland, 2014.

Noffsinger, Robert. "The Body as Commodity: Posthuman Tendencies in Bef's *Gel azul*." *Latin American Speculative Fiction*. Ed. Debra Ann Castillo and Liliana Colanzi. Spec. issue of *Paradoxa*, no. 30 (2018): 255–68.

Novo, Salvador. *La estátua de sal*. Prólogo de Carlos Monsiváis. Mexico City: Consejo Nacional para la Cultura y las artes, 1998.

Nunes, Jovane. *A escrava Isaura e o vampiro.* São Paulo: Lua de Papel, 2010.

Olímpio, Domingos. *Luzia-Homem.* 1902. São Paulo: Moderna, 1985.

Olvera, Carlos. *Mejicanos en el espacio.* 1968. Toluca: TunAstral, 2005.

Orbaugh, Sharalyn. "Sex and the Single Cyborg: Japanese Popular Culture Experiments in Subjectivity." *Science Fiction Studies* 29.3 (2002): 436–52.

Ortiz, Fernando. *Contrapunteo cubano: Tabaco y azúcar.* Caracas: Ayacucho, 1978.

Ortiz de Montellano, Bernardo. "Cinq heures sans coeur." *Cinco horas sin corazón: Entresueños.* México: Letras de México, 1940. 113–34.

Pacheco, José Emilio. "No perdura." *La sangre de Medusa y otros cuentos marginales.* Mexico City: Ed. Era, 1990. 85–87.

Padilla, Ignacio. "Las furias de Menlo Park." *Androide y las quimeras.* Madrid: Páginas de Espuma, 2009. 15–23.

Paiva, Marcelo Rubens. *Blecaute.* São Paulo: Brasiliense, 1986.

Palmer, Paulina. "The Lesbian Gothic: Genre, Transformation, Transgression." *Gothic Studies* 6 (May 2004): 118–30.

Pascal, H. [Juan Manuel García Junco]. *Fuego para los dioses.* Mexico City: Ed. Etoile, 1998.

Payne, Judith A., and Earl E. Fitz. *Ambiguity and Gender in the New Novel of Brazil and Spanish America: A Comparative Assessment.* Iowa City: U of Iowa P, 1993.

Paz, Octavio. *El laberinto de la soledad.* 1950. México: Fondo de Cultura Económica, 2004.

———. "La máscara y la transparencia." *Carlos Fuentes: Cuerpos y ofrendas.* Madrid: Alianza Editorial, 1981. 7–15.

Pearson, Wendy Gay. "Alien Cryptographies: The View from Queer." *Queer Universes: Sexualities in Science Fiction.* Ed. Wendy Gay Pearson, Veronica Hollinger, and Joan Gordon. Liverpool, UK: Liverpool UP, 2010. 14–38.

Pellegrini, Tânia. "Veríssimo: A inesperada subversão." *Gavetas vazias: Ficção política nos anos 70.* São Carlos, SP: UFSCar, 1996. 61–120.

Penglase, Ben. *Living with Insecurity in a Brazilia Favela: Urban Violence and Daily Life.* New Brunswick, NJ: Rutgers UP, 2014.

Pereira, Carla Cristina. See Lodi-Ribeiro, Gerson.

Pérez, Genaro. "La configuración de elementos góticos en 'Constancia,' *Aura* y 'Tlactocatzine, del jardín de Flandes.'" *Hispania* 80.1 (1997): 9–20.

Perez, Hiram. "Alma Latina: The American Hemisphere's Racial Melodramas." *Scholar and Feminist Online* 7.2 (2009). sfonline.barnard.edu/africana/print_perez.htm.

Perlongher, Néstor. *Prosa plebeya: Ensayos, 1980–1992.* Buenos Aires: Colihue, 1997.

Picatto, Pablo. "'Such a Strong Need': Sexuality and Violence in Belem Prison." *Gender, Sexuality, and Power in Latin America since Independence.* Ed. William E. French and Katherine Elaine Bliss. Lanham, MD: Rowman & Littlefield, 2007. 87–108.

Piletti, Nelson. *História do Brasil.* São Paulo: Ática, 1991.

Pirott-Quintero, Laura. "Strategic Hybridity in Carmen Boullosa's *Duerme*." *Ciberletras: Revista de crítica literaria y de cultura* 5 (2002): 1–8.

Poe, Karen. *Eros pervertido: La novela decadente en el modernism americano*. Madrid: Biblioteca Nueva, 2010.

Pons, María Cristina. "La mujer de la vida artificial: El tercer actor y el 'efecto travesti' de un imaginario decolonizador en *Duerme* de Carmen Boullosa." *Revista de literatura mexicana contemporánea* 24.74 (2018): 54–68.

Porcayo, Gerardo Horacio. *La primera calle de la soledad*. Mexico City: Fondo Adentro, 1993.

Potter, Sara Anne. "Disturbing Muses: Gender, Technology and Resistance in Mexican Avant-Garde Cultures." Diss., U of Washington, St. Louis, 2013.

Prado, Antônio Arnoni. *Lima Barreto: Literatura comentada*. Sao Paulo: Nova Cultural, 1988.

Prado, Paulo. *Retrato do Brasil: Ensaio sóbre a tristeza brasileira*. 1928. Rio de Janeiro: José Olympio, 1962.

Price, Brian L. *The Cult of Defeat in Mexico's Historical Fiction: Failure, Trauma, and Loss*. New York: Palgrave Macmillan, 2012.

Queiroz, Dinah Silveira de. "O carioca." *Comba Malina: Ficção científica*. 1960. Rio de Janeiro: Laudes, 1969. 179–204.

Queluz, Gilson Leandro, and Marilda Lopes Pinheiro Queluz. "*Kalum*: Utopia e modernismo conservador." *E-topia: Revista electronica de estudos sobre a utopia* 3 (2005): 1–7.

Rábago Palafox, Gabriela. "Primera comunión." *La voz de la sangre*. Toluca: Instituto Mexiquense, 1990. 21–28.

———. *La voz de la sangre*. Toluca: Instituto Mexiquense, 1990.

Rama, Ángel. *Transculturación narrativa en América Latina*. Mexico, DF: Siglo XXI, 1982.

Ramos, Samuel. *El perfil del hombre y la cultura en México*. 1934. Mexico City: UNAM, 1963.

Rebolledo, Efrén. "Salamandra." *Obras reunidas*. 1919. Ed. Benjamín Rocha. Mexico City: Oceano de México, 2004. 260–78.

Ribeiro, Darcy. *Os brasileiros: 1 teoria do Brasil*. Petrópolis: Editora Vozes, 1978.

Rieder, John. *Colonialism and the Emergence of Science Fiction*. Middletown, CT: Wesleyan UP, 2008.

Riley, Brandon. "The E-Dead: Zombies in the Digital Age." *Generation Zombie: Essays on the Living Dead in Modern Culture*. Ed. Stephanie Boluk and Wylie Lenz. Jefferson, NC: McFarland, 2011. 194–205.

Rio, João do. "O bebê de tarlatana rosa." 1910. *Histórias de gente alegre*. Ed. João Carlos Rodrigues. Rio de Janeiro: José Olympio, 1981. 56–61.

———. "História da gente alegre." 1910. *Histórias de gente alegre*. Ed. João Carlos Rodrigues. Rio de Janeiro: José Olympio, 1981. 49–55.

Ripoll Melo, Juan Vicente. "Mi velorio." *La noche alucinada*. N.p.: EFSA Editora, 1956. 23–34.

Rivera, José Eustacio. *La vorágine*. 1925. Mexico City: Editorial Porrúa, 1994.

Rojas Hernández, Arturo Céasar [Kalar Sailendra]. *Xxÿëröddny, donde el gran sueño se enraiza*. Mexico City: Panfleto y Pantomima, 1984.

Rojo, Pepe. "Ruido gris." 1996. *Los viajeros: 25 años de ciencia ficción mexicana*. Ed. Bernardo Fernández (Bef). Mexico City: SM de Ediciones, 2010. 98–129.

Rubião, Murilo. "Petúnia." *O convidado*. 1969. São Paulo: Quíron, 1974. 13–23.

———. "O pirotécnico Zacarias." 1943. *O pirotécnico Zacarias*. São Paulo: Ática, 1985. 12–19.

Rüsche, Ana, and Elton Aliandro Furlanetto. "Cultura e política nos anos 2010: Anseios e impasses na obra de Aline Valek e Lady Sybylla." *Abusões* 7.7 (2018): 253–91.

Sailendra, Kalar [Arturo César Rojas Hernández]. *Xxÿëröddny, donde el gran sueño se enraiza*. Mexico City: Panfleto y Pantomima, 1984.

Sánchez Prado, Ignacio. "*Amores perros*: Exotic Violence and Neoliberal Fear." *Journal of Latin American Cultural Studies: Travesia* 15.1 (2006): 39–57.

———. "El mestizaje en el corazón de la utopía: La raza cósmica entre Aztlán y América Latina." *Revista de estudios hispánicos canadienses* 33.2 (2009): 381–404.

———. "War and the Neoliberal Condition: Death and Vulnerability in Contemporary Mexico." Invited lecture by the Center for the Humanities and the Public Sphere. 15 Sept. 2016, University of Florida, Gainesville, FL.

Santiago, Silviano. "O entre-lugar do discurso latino-americano." *Uma literatura nos trópicos: Ensaios sobre dependência cultural*. N.p.: Editora Perspectiva, 1978. 11–28.

———. "O homossexual astucioso." *Brasil/Brazil* 13.23 (2000): 7–18.

Sardà, Alejandra, Rosa María Posa Guinea, and Verónica Villalba. "Lesbianas en América Latina: De la inexistencia a la visibilidad." *Mujeres en la red: Periódico feminista*, Jan. 2006. www.mujeresenred.net/spip.php?article1349.

Sarduy, Severo. "The Baroque and the Neo-Baroque." Trans. Mary G. Berg. *Latin America in Its Literature*. Ed. César Fernández Moreno, Julio Ortega, and Ivan A. Schulman. New York: Holmes and Meier, 1980. 115–32.

———. "La simulación." *Ensayos generales sobre el barroco*. Buenos Aires: Fondo de Cultura Económica. 1987. 53–146

Schaffer, Talia. "'A Wilde Desire Took Me': The Homoerotic History of Dracula." *Dracula*. By Bram Stoker. 1897. Ed. Nina Auerbach and David J. Skal. New York: Norton, 1997. 470–82.

Sclofsky, Sebastian. "Policing in Two Cities: From Necropolitical Governance to Imagined Communities." *Journal of Social Justice* 6 (2016): 1–24.

Schwarcz, Lilia Moritz. *O espectáculo das raças: Cientistas, instituições e questão racial no Brasil 1870–1930*. São Paulo: Companhia das Letras, 1993.

Schwarz, Roberto. *Ao vencedor as batatas*. São Paulo: Duas Cidades, 1977.

———. "Misplaced Ideas." *Comparative Civilizations Review* 5.5 (1980): 33–51. scholarsarchive.byu.edu/ccr/vol5/iss5/3.

Seabrook, William. ". . . Dead Men Working in the Cane Fields." *The Magic Island*. New York: Harcourt, Brace and Company, 1929. 92–103.

Sedgwick, Eve Kosofsky. *Between Men: English Literature and Male Homosocial Desire*. New York: Columbia UP, 1985.

Seed, David. *Science Fiction: A Very Short Introduction*. Oxford: Oxford UP, 2011.

Seltzer, David. *Bodies and Machines*. 1992. New York: Routledge, 2014.

Serna, Enrique. "Tía Nela." *El orgasmógrafo*. Mexico City: Seix Barral, 2010. 191–203.

Serra, M. V., and Antonio Edmilson Martins Rodrigues. Introduction. *A alma encantadora das ruas*. By João do Rio. Ed. and trans. Mark Carlyon. Rio de Janeiro: Cidade Viva, 2010. 28–41.

Serrano, Carmen. "Mapping the Zombie: Diego Velazquez Betancourt's Newfangled Zombie in *La noche que asolaron Tokio*." *Romance Notes* 53.8 (2018): 461–72.

———. "Revamping Dracula on the Mexican Silver Screen: Fernando Méndez's *El vampiro*." *Vampires and Zombies: Transcultural Migrations and Transnational Interpretations*. Ed. Dorothea Fischer-Hornung and Monika Mueller. Jackson, MS: UP of Mississippi, 2016. 149–67.

Serrato Córdova, José Eduardo. "El imaginario gótico en dos autores mexicanos: Emilio González y Ernesto de la Peña." *Isla flotante* 5 (2013): 27–44.

Shildrick, Margrit. *Embodying the Monster: Encounters with the Vulnerable Self*. London: Sage, 2002.

Showalter, Elaine. *Sexual Anarchy: Gender and Culture at the Fin de Siècle*. New York: Viking, 1990.

Sifuentes, Gerardo, and Bernardo Fernández (Bef). "(e)." *Visiones periféricas: Antología de ciencia ficción mexicana*. Ed. Miguel Ángel Fernández Delgado. Mexico City: Lumen, 2001. 173–86.

Sifuentes-Jáuregui, Ben. *Transvestism, Masculinity and Latin American Literature: Genders Share Flesh*. New York: Palgrave, 2002.

Simão Branco, Marcello, and César Silva. "Personalidades do ano: Entrevista com Giulia Moon e Martha Argel." *Anuário brasileiro de literatura fantástica 2006*. Ed. Marcello Simão Branco and César Silva. São Bernardo do Campo, SP: Hiperespaço, 2007. 57–67.

Skidmore, Thomas E. *Black into White: Race and Nationality in Brazilian Thought*. 1974. Durham, NC: Duke UP, 1993.

Skidmore, Thomas E., and Peter H. Smith. *Modern Latin America*. Oxford: Oxford UP, 1992.

Slater, Candace. *Entangled Edens: Visions of the Amazon*. Berkeley: U of California P, 2002.

Sommer, Doris. *Foundational Fictions: The National Romances of Latin America*. Berkeley, CA: U of California P, 1991.

Souza, Márcio. *A ordem do dia*. Rio de Janeiro: Marco Zero, 1983.

———. *The Order of the Day*. Trans. Thomas Colchie. New York: Avon Books, 1986.

Stepan, Alfred. *The State and Society: Peru in Comparative Perspective.* Princeton, NJ: Princeton UP, 1978.

Stepan, Nancy. *"The Hour of Eugenics": Race, Gender, and Nation in Latin America.* New York: Cornell UP, 1991.

Stevenson, Robert Louis. *Strange Case of Dr. Jekyll and Mr. Hyde.* 1886. Lincoln: U of Nebraska P, 1990.

Stoker, Bram. *Dracula.* 1897. Ed. Nina Auerbach and David J. Skal. New York: Norton, 1997.

Suárez y López Guazo, Laura Luz. *Eugenesia y racismo en México.* Mexico City: UNAM, 2005.

Subero, Gustavo. "Zé do Caixão and the Queering of Monstrosity in Brazil." *Gender and Sexuality in Latin American Horror Cinema: Embodiments of Evil.* London: Palgrave, 2016. 39–71.

Süssekind, Flora. *Tal Brasil, qual romance? Uma ideologia estética e sua história: O naturalismo.* Rio de Janeiro: Achiamé, 1984.

Suvin, Darko. *Metamorphoses of Science Fiction: On the Poetics and History of a Literary Genre.* New Haven, CT: Yale UP, 1979.

Tarazona, Diana. *El animal sobre la piedra.* Mexico City: Almadía, 2008.

Tavares, Braulio. "Stories of the Will-Happen: Science Fiction in Brazil." *Foundation* 77 (1999): 84–91.

———, ed. *Páginas de sombra: Contos fantásticos brasileiros.* Rio de Janeiro: Casa da Palavra, 2003.

Tavares, Braulio, and Roberto de Sousa Causo. "Brazil." *The Encyclopedia of Science Fiction* 25 Apr. 2016 Ed. John Clute, David Langford, Peter Nicholls, and Graham Sleight Gollancz. www.sf-encyclopedia.com/entry/brazil.

Taylor, Claire. "Geographical and Corporeal Transformations in Carmen Boullosa's *Duerme.*" *Bulletin of Hispanic Studies* 83.3 (2006): 225–39.

Telles, Lygia Fagundes. "Seminário dos ratos." *Seminário dos ratos.* 1977. Rio de Janeiro: José Olympio, 1980. 116–26.

———. "Tigrela." *Seminário dos ratos.* 1977. Rio de Janeiro: José Olympio, 1980. 22–27.

Tichi, Cecilia. *Shifting Gears: Technology, Literature, Culture in Modernist America.* Chapel Hill: U of North Carolina P, 1987.

Tiptree, James, Jr. [Alice Sheldon]. "The Women Men Don't See." *Her Smoke Rose Up Forever.* 1973. Introd. Michael Swanick. San Francisco: Tachyon, 2004. 115–44.

Todorov, Tzvetan. *The Fantastic: A Structural Approach to the Literary Genre.* Trans. Richard Howard. Ithaca, NY: Cornell UP, 1975.

Tola de Habich, Fernando, and Angel Muñoz Fernández. "Alejandro Cuevas, 'El vampiro.'" *Cuento fantástico mexicano: Siglo XIX.* Mexico City, Factoría Ediciones, 2005. 289–302.

Trevisan, Dalton. *O vampiro de Curitiba.* 1965. Rio de Janeiro: Editora Record, 2008.

Trevisan, João Silvério. *Devassos no paraíso: A homossexualidade no Brasil, da colônia à atualidade.* 1986. Rio de Janeiro: Editora Record, 2000.

Trujillo Muñoz, Gabriel. *Biografías del futuro: La ciencia ficción mexicana y sus autores.* Mexicali, BC: Universidad Autónoma de Baja California, 2000.

———. *Espantapájaros.* Mexico City: Lectorum, 1999.

Twitchell, James B. *The Living Dead: A Study of the Vampire in Romantic Literature.* Durham, NC: Duke UP, 1981.

Tyson, Peter. "The Experiments." *Holocaust on Trial.* Nova Online, October 2000. www.pbs.org/wgbh/nova/holocaust/experiside.html.

Urquizo, Francisco L. *Mi tío Juan: Novela fantástica.* Mexico City: Claret, 1934.

Urzaiz, Eduardo. *Eugenia: Esbozo novelesco de costumbres futuras.* 1919. Introd. Carlos Peniche Ponce. Mexico City: UNAM, 2006.

Valek, Aline. *As águas-vivas não sabem de si.* Rio de Janeiro: Rocco, 2016.

Varley, John. "Picnic on Nearside." *The John Varley Reader: 30 Years of Short Fiction.* 1974. New York: Ace, 2004. 1–23.

———. "Options." *The John Varley Reader: 30 Years of Short Fiction.* 1979. New York: Ace, 2004. 409–36.

Vasconcelos, José. *La raza cósmica: Misión de la raza iberoamericana, Argentina y Brasil.* 1925. Mexico City: Espasa-Calpe, 1966.

Vejmelka, Marcel. "O Japão na literatura brasileira atual." *Estudos de literatura brasileira contemporânea* 43 (Jan-Jun 2014): 213–34.

Veiga, José J. *A hora dos ruminantes.* Rio de Janeiro: Civilização Brasileira, 1966.

———. *The Three Trials of Manirema.* Trans. by Pamela G. Bird. New York: Knopf, 1970.

Velázquez Betancourt, Diego. *La noche que asolaron Tokio.* Mexico City: Conaculta, 2013.

Veríssimo, Erico. *Incidente em Antares.* Porto Alegre: Globo, 1971.

Vianco, André. *Os sete.* 1999. São Paulo: Novo Século, 2003.

Vicente, Gil. *Auto da barca do inferno.* 1517. Porto Alegre: L&PM Pocket, 2005.

Vieira, Nelson H. "Closing the Gap between High and Low: Intimation on the Brazilian Novel of the Future." *Latin American Literary Review* 20.4 (1992): 109–19.

Villa, Andrea. "Contrapunteo neobarroco entre la literatura y las artes visuales de Clarice Lispector, Cristina Rivera Garza, Laura Restrepo, Lygia Clark, Graciela Iturbide y Doris Salcedo: Aproximación a un nuevo discurso estético en Latinoamérica." Diss., U of Florida, 2016.

Villiers de L'Isle Adam, Auguste. *Tomorrow's Eve.* Trans. Robert Martin Adams. Urbana: U of Illinois P, 2001.

Vint, Sherryl. "Abject Posthumanism: Neoliberalism, Biopolitics, and Zombies." *Zombie Theory: A Reader.* Ed. Sarah Juliet Lauro. Minneapolis: Minnesota UP, 2017. 171–81.

Vivanco, José Miguel. "A Voice for LBGT Rights Silenced in Brazil." Human Rights Watch, 24 January 2019. www.hrw.org/news/2019/01/24/voice-lgbt-rights-silenced-brazil.

Webb, Jen, and Samuel Byrnand. "Some Kind of Virus: The Zombie as Body and as Trope." *Zombie Theory: A Reader.* Ed. Sarah Juliet Lauro. Minneapolis: Minnesota UP, 2017. 111–23.

Wegner, Phillip. *Life between Two Deaths, 1989–2001: U.S. Culture in the Long Nineties.* Durham, NC: Duke UP, 2009.

Wiarda, Howard J., and Harvey F. Kline. *Latin American Politics and Development.* 8th ed. Boulder, CO: Westview, 2013.

White, Patricia. "Female Spectator, Lesbian Specter: *The Haunting.*" *The Horror Reader.* Ed. Ken Gelder. London: Routledge, 2000. 210–22.

Wiener, Norbert. *The Human Use of Human Beings: Cybernetics and Society.* Boston: Houghton Mifflin, 1950.

Wilde, Oscar. *The Picture of Dorian Gray.* 1890. New York: U of Oxford P, 1974.

Yanielli, Joseph. "Your Slavery Footprint." *Digital Histories@Yale,* 14 Oct. 2011. www.hastac.org/blogs/jyannielli/2011/10/14/your-slavery-footprint

Zárate, José Luis. *La ruta del hielo y la sal.* Mexico City: Vid, 1998.

———. "El viajero." 1987. *Los viajeros: 25 años de ciencia ficción mexicana.* Ed. Bernardo Fernández (Bef). Mexico City: SM Ediciones, 2010. 51–66.

———. *Xanto: Novalucha libre.* Mexico City: Planeta, 1994.

Zola, Émile. "The Experimental Novel." *The Experimental Novel and Other Essays.* Trans. Belle M. Sherman. New York: Haskell House, 1964. 644–55.

INDEX

"Abaporu" (Amaral), 184n3
abertura, 133
abjection, 50, 56, 91
    functions and horror, 183n33
    transformations, 21, 101
Abreu, Caio Fernando, 22, 180n23, 184n1
"Academias de Sião, As" (Machado),
    76–79
access to information, 59
addictions
    cyber, 134, 183n34
    drugs, 57
    news, 183n34
    See also zombies: consumer;
        vampires
advertising
    in Arreola, 41, 43
    in Becerra Acosta, 125
    and consumer durables, 39
    and consumption, 43, 125
Afonso Reyes e o Brasil (Ellison), 173n14,
    174n21
Afro-Brazilian cultural practices,
    180n21. See also Black
Afrodescendant. See Black
Agamben, Giorgio

and bare life, 17–18
and docile bodies, 195
and homo sacer, 18
and profanation, 123–24, 134, 191n12,
    195n12
and sovereign power, 145–46
    See also biopolitics
Águas-vivas não sabem de si, As (Valek), 102
Aguilera, Pedro, 167
AIDS, 51, 110, 180n23, 184n5
alchemy, 32, 114. See also arcane sci-
    ences; pseudo-sciences
Alemán, Miguel, 145
Alencar, José de, 170n3
algorithms
    in cybernetic systems, 65
    in necropolitics, 2
aliens, 95, 96–97, 104
    and men, 98, 188n26
    and queerness, 185n7
Alma encantadora das ruas, A (Rio),
    178n9
"Altar dos nossos corações, O" (Calado),
    55, 182n32
alternate modernity
    and the body, 21–25, 28

alternate modernity (*continued*)
and Echeverría, 45, 108, 116
and history, 53, 181n28
in speculative fiction, 167–68
Amaral, Tarsila do, 184n3
Amazon, 76
as collective unconscious, 80, 104
degradation of, 96
as lost civilization, 81
as setting, 79–82, 95–98
and women warriors, 81–83
Amazônia misteriosa A, (Cruls), 79–83,
186n9, 186n11
Amazons to Zombies (Braham), 20, 21,
141, 164
*Ambiguity and Gender in the New Nov-
els of Brazil and Spanish America*
(Payne and Fitz), 173n14
amborg, 186n11
Amor, Pitta, 43–44
Amores perros (Iñárritu), 182n30
Amor vampiro (Moon), 195n13
Anderson, Benedict, 6
Andrade, Carlos Drummond de, 24,
126–27
Andrade, Mário de, 186n14
Andrade, Oswald de, 13, 71, 79, 108,
172n13, 175n27, 184n3
androgynous, 78, 93, 99–100, 189n27.
See also queerness
Androides y las quimeras, Los (Padilla),
181n27
animals
and experiments, 82, 158–60
and Haraway, 45
and rights as inteligent species, 160
women's identification with, 23,
98–104
Animal sobre la piedra, El (Tarazona),
100–102
antigens, 137. *See also* immunity
antropofagia
cultural, 80

antropofagia (*continued*)
and cultural immunity, 18, 108, 127–
28, 135–36, 138
and devouring of culture, 13, 28, 67,
72, 80, 103nn2–3
and queerness, 23, 71–73, 103–4, 184n4
and third space, 28–29
and zombies, 108–9
See also cannibal; *codigofagia*;
zombies
"Anuncio" (Arreola), 41–43
Ao vencedor as batatas (Schwarz), 31
Aponte Alsina, Marta, 189n29
Arata, Stephen, 140, 156, 159, 188n21
Arboleda-Ríos, Paola, 184n1
arcane sciences, 92, 94
and queering, 87–89, 95
See also occult
Arenas, Fernando, 48
Argel, Martha, 195n13
Argentina
compared to Mexico and Brazil,
169n2
and eugenics, 173n19
and export economy, 29
and science fiction, 8, 177n5, 193n6
and stock market novel, 190n3
Arias, Desmond, 181n25
Arias, Rosario, 150
Aridjis, Homero, 24, 120–23, 191n9
Arreola, Juan José, 19, 22, 41–44, 67–68
"Ascensão e queda de Robhéa, manequim
e robô, A" (Abreu), 48–50, 67–68
asexual
and female behavior, 44, 47
reproduction, 8, 99
assembly lines, 40, 125. *See also*
consumerism
"Assessor para assuntos fúnebres"
(Lodi-Ribeiro), 155–56
assimilationist policies, 7, 10, 14, 71,
156, 172n13. See also Black: racial
whitening; *mestizaje/mestiçagem*

Atahualpa, 81
Ateneo de la Juventud, 174n21
Atlantis, 81. *See also* lost civilizations
Atteberry, Brian, 74
authoritarianism, 2, 24, 46–47
    and biopolitics, 18
    and crackdowns in Mexico and Bra-
        zil, 47, 129–30
    and liberation from, 185n8
    regimes in fiction, 48, 130, 192n17
auto-amputation, 58, 183n35–36
Auto da barca do inferno, 131
autoimmunity, 17, 175n30
    as disease, 150
    and gothic mansion, 146, 150–53
    as state corruption, 17, 132, 135
automatons, 32, 37, 51–53, 181n26
autopoesis, 180n18. *See also* reflexivity
Autopoeisis and Cognition (Maturana
    and Varela), 180n18
avatars, 57, 59–60, 64, 67
Avelar, Idelber, 49–51
Aventuras do vampiro de Palmares (Lodi-
    Ribeiro), 154, 157
Avilés, Edgar Omar, 185n5
Ayotzinapa, State of Guerrero, 129
Azevedo, Aluisio, 115, 188n23
Aztlán, 9

"Baby H. P." (Arreola), 41
Bacigalupi, Paolo, 66
Baden, Nancy, 192n15
Bailey, Stanley, 180n21
Barcia, Jacques, 180n20
Baring Crisis, 190n3
baroque art, 1–2
baroque ethos, 11–12, 27–28
    and cyborgs, 22–24, 56, 69, 167
    and queer, 15, 21, 71–72, 74, 76, 102,
        104–5
    and vampires, 135, 138, 143, 148–49, 154
    and zombies, 17, 108, 128
Barrera-Barrios, 99

Bartra, Roger, 172n13
Bates, Henry Walter, 81
bats, 154, 158–59, 163
Battle of the Sexes in Science Fiction, The
    (Labalestier), 74
"Bebê de tarlatana rosa, O" (Rio), 35
Becerra Acosta, Manuel, 24, 125–26
Beckman, Ericka, 22, 29, 34, 35, 55,
    112–13, 122, 178n7, 182n31, 190n3
Bef, 22, 57, 60, 67, 183n36, 185n5, 192n18.
    See also Fernández, Bernardo
Beisdecker, Dave, 110
Bell, Andrea L., 176n32
Bellatín, Mario, 184n5
belle époque, 76
Bermúdez, Maria Elvira, 189n1
Bernard, Claude, 173n17
Besançon, Guy, 150
Between Men (Sedgwick), 87
Biehl, João, 165
binaries, 166–67
    and baroque, 68
    breakdown of, 12
    and gender, 60, 85, 89, 95
    and social order, 78, 167
    and space in-between, 28, 73–74,
        86–87, 94
    and zombies, 107, 124
    See also hybridity
Biografias del futuro (Trujillo Muñoz),
    75, 176n34
biopolitics, 3, 7, 13, 167–68, 169n1
    affirmative, 7, 17, 25, 165, 167, 176n31
    and Agamben, Giorgio, 17–18,
        195n12
    and Esposito, 16–17, 127–28, 130, 134,
        136–37
    Foucault, 3, 17, 42, 170n5
    and immunity, 16–17, 24–25, 114–15
    and socio-political control, 42, 57, 134
    and work force, 42, 57, 134, 191n12
    See also necropolitics
biopower, 42–43, 165

bios, 17–18, 25, 136, 154
  as phantasm of life and death, 159–60
  See also Agamben, Giorgio; *zoê*
Bios: Biopolitics and Philosophy
  (Esposito), 136, 143, 144
biotechnology, 125–26, 158
Birds, The, 110
birth, 84
  and becoming, 160–61
  male incubators, 189n27
  and mestizo nation, 6
  stillbirth, 102
  See also pregnancy; reproduction
bisexuality, 48, 62, 94. *See also*
  queerness
Black, 4, 9, 13
  authors, 36, 115, 178n10, 187n14
  Brazilian culture, 9, 174n20, 180n21
  characters, 96, 103, 156–57, 188n23
  population and colonization, 71, 155
  racial whitening, 7, 10
  role in *mestizaje/mestiçagem*, 9
  women, 153, 156–57
  See also slavery
Black into White (Skidmore), 7
Blecaute (Paiva), 191n8
Bliss, Katherine Elaine, 15
blood, 7
  as antigen, 138
  bad, 145–46, 154, 158–59, 166
  cold blooded, 23, 99, 102
  and degeneration, 136–37
  in *mestizaje*, 85–86, 156
  and vampires, 137, 138, 142, 163,
    165–66, 193n2
  See also autoimmunity
Bodies and Machines (Seltzer), 179n14
Bodies that Matter (Butler), 50, 96–97
body, 2–4, 13
  and cultural correlates of, 86, 170n3,
    179n14
  queer, 23, 50, 62, 73–76, 80, 83, 88, 91,
    96–97, 100, 103

body (*continued*)
  reproductive, 6, 8, 43, 60, 75, 84, 95,
    99–100, 105, 137, 163
  in speculative fiction, 19–20, 27, 56
  and vampires, 142–43, 145, 147, 149,
    151–52, 156, 158
  and Western tradition, 2, 30, 76–77
  and zombies, 114, 116, 119–20, 122, 131
  See also abjection; biopolitics
body politic, 3, 6, 46, 167
bokor, 120, 126
Bolsonaro, Jair, 180n22
Boluk, Stephanie, 110, 111
Bom-Crioulo (Caminha), 188n23
border relations Mexico, 160–61
Borges, Dain, 115, 173n19, 178n10
Borges, Jorge Luis, 52
Borinsky, Alice, 177n5
Botting, Fred, 20, 21
Boullosa, Carmen, 23, 31, 83–86, 105,
    177n5
Boyle, Danny, 110
bracero program, 159
Braham, Persephone, 20, 21, 80, 141, 164
Branco, Marcello Simão, 196n18
Brasileiros, Os (Ribeiro), 172n13
*Brazilian Science Fiction: Cultural Myths
    and Nationhood* (Ginway), 1
Brescia, Pablo, 41
Brontë, Charlotte, 194n9
Brontë, Emily, 139
Brooks, Max, 110
Brown, Jacob C., 138, 141, 153, 156–58,
    195n12
Brown, J. Andrew, 21, 31, 177n5
Buarque de Holanda, Sérgio, 172n13
Buffy the Vampire Slayer, 161
Burton, Tara Isabella, 179n15
Butler, Judith, 15, 60, 75, 85, 91, 96–67, 166
Byrnand, Samuel, 121, 124, 125
Byronic hero, 186n11

"Café, café" (Lobato), 115–16

Calado, Ivanir, 182n32

calavera, 117–18

zombie calavera, 118–19

See also skulls

Caldeira, Teresa, 173n16

Calderón, Felipe, 122

Calogero, Stephen, 173n18

Campbell, Timothy, 123, 138, 165, 176n31, 193n3, 195n12

Campos, Haroldo de, 71, 191n11

Campos, Victoria E., 173n15

Canclininéstor García, 9

Cañedo, Diego, 19

cannibal

and cultural resistance, 108, 123

as liberated from work, 185n8

and "Manifesto antropófago," 79, 172n13

monstrous practices of, 80, 136

and zombie, 109

See also Amazon; Andrade, Oswald de; zombies

Cano, Luis, 37, 179n12, 187n16

Canudos, 191n7

Capital Fictions (Beckman), 22, 29

capitalism, 10–13, 27–28, 30, 167, 175n27

and vampires, 140

and zombies, 110–11, 115, 124, 134

See also colonialism; neoliberalism

"Capitão dos Geraes, O" (Lodi-Ribeiro), 157

"Capitão Mendonça, O" (Machado), 31, 178n8

capoeira, 174n20, 180n21

Cardona Peña, Alfredo, 22, 193n2

Cardoso, Lúcio, 25, 136, 146

Carelli, Mário, 149

"Carioca, O" (Queiroz), 44–46

Carlota, Empress of Mexico, 144

Carmilla (LeFanu), 195n11

Carnavais, malandros e heróis (DaMatta), 172n13

Carneiro, André, 185n5

carnival, 34–36, 49, 72, 174n20

Caruth, Cathy, 97

Carvajal, Gaspar de, 80

Casa grande e senzala (Freyre), 9

Casanova-Vizcaíno, Sandra, 20, 141

Castellanos, Rosario, 43

Castera, Pedro, 22, 36–37, 179n12, 187n16

caste system in Mexico, 171n8

Castillos en la letra (Moussong), 193n2

castration complex, 177–78n6

Catholic Church, 87, 140, 170n3, 193n2

Catrina, La, 117

Causo, Roberto de Sousa de, 21, 23, 60–62, 80, 90–92, 95, 105, 114, 176n34, 186n9

cellular phones, 127

censorship, 19, 130, 132, 192n15

Certeau, Michel de, 3, 171n10

Chacek, Karen, 24, 127, 133–34, 192n16

Chapo, El (Joaquín Guzmán Loera), 122, 191n10

Charlot, Jean, 118

charro, 118

Chávez, Carlos, 174n20

Christ, 3, 153, 163

chupacabras, 159

Cidades mortas (Lobato), 115

Cielos en la tierra (Boullosa), 31, 177n5

científicos, los, 7. See also positivism

"Cinq heures sans coeur" (Ortiz de Montellano), 124

Cirne, Roberta, 190n6

City of God (Meirelles and Lund), 182n30

Ciudad de los zombis, La, 120–22

Clark, Lygia, 184n3

class, 2, 4, 14, 36, 41, 103, 108–9, 121, 164

middle, 40, 91

struggle, 103, 115, 118

upper, 34, 115–16, 121, 151

working, 5, 27, 119–20, 165

cloning, 95, 96

Clute, John, 176n32

codigofagia, 12–13, 21, 28, 34–35, 43, 59, 62,
    67–69, 71, 97, 106

Coelho Neto, 23, 90–95, 105, 178n9,
    187n16

coffee, 4, 30, 112–13, 115–17
    boom, 178n8
    and export economy, 29, 33–34,
        112–13

cognitive estrangement, 20, 66

Cohen, Jeffrey Jerome, 104

Cold War, 109, 129

Collor de Melo, Fernando, 54

colonialism, 4, 12–13
    and colonizer-colonized, 28–29
    and enslaved, 3, 33, 178n7
    and gender, 23, 83–86
    and indigenous, 79–83
    and labor, 56, 115, 171n11
    and *mestizaje/mestiçagem*, 6, 9–10, 14
    and subaltern population, 4, 10, 108
    and vampires, 25, 136, 140–41, 157,
        154–58, 162, 166

Coluna Prestes, 191n7

Comando Vermelho, 182n32

Comaroff, Jean and John, 182n31

commodities, 27, 108, 112, 122
    and booms, 122
    and fetishism, 29, 33–34, 113, 115,
        178n7, 179n13
    and madness, 178n8
    and slavery, 178nn7–8
    See also coffee; economics

communism, 129. *See also* Cold War;
    dictatorship

communitas, 131, 138. *See also* Esposito,
    Roberto

comparative studies of Mexico and
    Brazil, 5

Comte, Auguste, 6–7, 94, 173n18. *See
    also* positivism

CONACYT, 19

Connor, Anne, 185n5

conquest, 3, 10, 14, 81–86, 90, 104–5. *See
    also* colonialism

Conrad, Joseph, 96

consumerism, 5, 22
    and cyborgs, 39–43, 48–49
    and zombies, 24, 109–11, 114, 123–26
    See also consumption

Consuming Youth (Latham), 195n14

consumption, 5, 27, 30, 36, 40–41
    and fetishism, 179n13
    and import substitution, 39
    internal, 26-27, 40, 45
    and zombies, 124–25
    See also commodities

Contos fantásticos (Machado), 178n9

Cornejo Polar, Antonio, 9

corporate state, 3, 170n3, 170n6, 188n25

corporations, 23, 39, 54, 57
    and the body, 57, 109
    See also neoliberalism

Corral Rodríguez, Fortino, 148

corruption, 2, 5–6, 56–57, 166
    as auto-immune response, 132, 145,
        151
    in Brazil, 61, 120, 132
    in Mexico, 61, 118, 120, 122–23, 145

Cortés, Hernán, 187n15

Cortiço, O (Azevedo), 115, 188n23

Cortines, Adolfo Ruiz, 146

Cosmic Race, The (Vasconcelos), 9–10,
    138, 172n13

Cosmos Latinos (Bell and Molina
    Gavilán), 176n32

counter-immunity, 17–18, 24
    and vampires, 138, 141, 159, 162, 164
    and zombies, 128, 131, 133–35
    See also immunity

couplings, 30, 44, 68

Crack generation, 181n27

Craft, Christopher, 193n5

Crafting the World (Love), 190n2

Crawford, Heide, 139

Creed, Barbara, 100, 140, 151

crime, 2, 5
    and cyberpunk, 58–63
    and fiction, 57, 67
    and homosexuality, 14–15, 87, 91,
        175nn28–29
    and networks, 55–57
    and the state, 169n1
    and urbanization, 54
    and zombies, 120–21
Crônica da casa assassinada, A
    (Cardoso), 136, 146, 148–53,
        194n8–9
Cronos (del Toro), 141, 164
cross-species
    breeding, 103, 186n11, 186n13
    experimentation, 81–82
    and identification, 100–104
    See also oviparity
Cruls, Gastão, 23, 79–82, 86, 186n11
Crump, Marty, 189n28
Cuarenta y uno, Los (Castrejón), 175n29
Cuenca, João Paulo, 23, 60, 63, 65–68,
    183n37
Cuento fantástico mexicano: Siglo XIX
    (Tola de Habich), 142
Cuevas, Alejandro, 24, 136, 142
*Cult of Defeat in Mexico's Historical
    Fiction* (Price), 52, 172n12
Cummanká (Del Picchia), 188n25
cybernetics, 27, 29–30, 40, 53, 177n2
    first phase, 42, 48, 52, 180n17
    second phase, 44, 53
    third stage, 64-65
    Wienernorbert, 30, 40, 180n17
    See also couplings; Hayles, Kather-
        ine; signifier: flickering
cyberpunk, 19, 55–62, 180n20, 182n32
    American, 53–54, 55
    Latin American, 55–56, 183n34,
        183n36
cyberspace, 23, 30, 53, 56
    and access to, 59
    in cyberpunk, 53, 55, 57–60

cyberspace (*continued*)
    and disembodied capital, 54
    See also auto-amputation; cyberpunk
cyborgs, 12, 26, 27, 62
    and animals, 45–46
    female, 27, 177n5, 179n15, 180n20
    and horror, 177n35
    and posthuman, 27, 31, 50, 52–53, 59,
        63, 65–66, 69
    three cybernetic stages of develop-
        ment, 26-27
Cyborgs in Latin America (Brown, A.),
    21, 177n5

Dalton, David, 18, 21, 108, 127–28, 133
DaMatta, Roberto, 14, 53, 171n9, 172n13
Darwin, Charles, 187n16. *See also* social
    Darwinism
Dávila, Amparo, 25, 137, 146–48, 152
Dawn of the Dead, 110
Day of the Dead, 110, 117, 174n20
De Castro, Juan E., 4, 10
De Cicco, Mark, 87, 89, 95
Decoding Gender in Science Fiction (Atte-
    berry), 74
degeneration, 7, 94, 115, 139, 173n19,
    178n10. *See also* eugenics;
    positivism
*Delmira Augustini, Sexual Seduction, and
    Vampiric Conquest* (Jrade), 141
Del Picchia, Paulo Menotti, 186n10,
    188n25
democracy, 6, 18
    in Brazil, 33, 46, 51, 120, 173n16
    disjunctive, 5, 173n16
    and dual state, 181n25
    in Mexico, 60, 118, 172n12, 188n25
    racial, 9–10, 158, 180n21
Dentes negros (Leones), 120
Derrida, Jacques, 53, 77–79, 182n20
Diário do hospício e Cemitério dos vivos
    (Lima Barreto), 190n4
Díaz, Porfirio, 7, 117, 171n7

Díaz Ordaz, Gustavo, 129
Dick, Philip K., 61
dictatorship, 19, 132–33
    in Argentina, 170n25
    in Brazil, 46, 48–51, 61, 68, 130–32, 133,
        182n2, 185n5, 191n8
    in Mexico, 47, 129, 191n9
    in Spanish American literature, 21,
        177n5
    See also authoritarianism
Discipline and Punish (Foucault), 64,
    179n16
Ditadura escancarada, A (Gaspari), 191n13
Do Androids Dream of Electric Sheep
    (Dick), 61
dolls, 22, 27, 181n27
    and cyborgs, 181n27
    and sex, 42, 63
Dom Pedro I, vampiro (Fonseca), 157,
    195n12
Dom Pedro II, 32–33, 191n14
donador de almas, El (Nervo), 88–91, 94,
    97, 179n15, 187n17
doubles, 150–52
    and the fantastic, 96–97
    and Freudian uncanny, 178n6
    and vampires, 163
    and zombies, 121
Dracula (Stoker), 91, 139–40, 140, 155, 162
drugs, 55–57, 183n34
    illegal, 96, 119–23, 158
    legalization of, 183n36
    War on, 123, 161, 169n12, 182n30,
        191n10
Duerme (Boullosa), 31, 83–86, 187n15
Du Maurier, Daphne, 150, 194n9
Dunbar, David L., 45
dystopia, 19, 48, 168, 184n5. See also
    dictatorship

"(e)" (Sifuentes and Fernández), 56
Echeverría, Bolívar, 10–13, 16, 22, 23,
    27–29, 45, 47, 56, 136

    and baroque ethos, 10–12, 27–28,
        174n24
    and codigofagia, 13, 28, 62, 69
    and colonization, 71
    and magical realism, 175n25
    and resistance, 74, 107, 108, 174n24
    and revolution, 175nn26–27
    and use value, consumerism, 179n13
economics, 5, 39, 54, 117
    and exports, 29, 34, 112–13, 190n3
    and import substitution, 41, 46
    and neoliberalism, 54–55, 57, 60, 122,
        134, 182n31
Edison, Thomas Alva, 179n15, 181n27
Edwards, Erika, 170n2
Edwards, Justin D., 20, 141
Eljaiek-Rodríguez, Gabriel Andrés, 140,
    193n6, 195n12
Ellison, Fred, 172–73n14
embodiment, 2, 13, 15
    freedom from, 56–57, 65
    and gender, 187n20
    and labor, 29
    in Latin America, 75
    as resistance, 68–69, 107
    and social belonging, 74
    See also body
Embodying the Monster: Encounters with
    the Vulnerable Self (Shildrick), 63
Embry, Karen, 110–11
Emergence of Latin American Science Fic-
    tion, The (Haywood Ferreira), 8,
    18, 21, 189n27
Encilhamento, 112, 190n3
Encyclopedia of Science Fiction, 176n32
entre-lugar, 73–74. See also Santiago, Sil-
    viano; space in-between
entropy, 40, 45, 179n14, 180n17. See also
    thermodynamics
environmental destruction, 96, 105
Erauso, Catalina de, 83
Eros decadente en el modernismo mexi-
    cano (Poe), 187n17

Escrava Isaura e o vampiro, A (Nunes), 195n13

Esfinge (Coelho Neto), 90–95, 188n24

Espantapájaros (Trujillo Muñoz), 137, 154, 158–61

Espectáculo das raças, O (Schwarcz), 173n19

Espíritus de la ciencia ficción, Los (Cano), 179n12, 187n16

Esposito, Roberto, 16, 25, 127, 134, 160
  autoimmune response, 17, 132, 135, 150, 152–53, 214n10
  on birth and becoming, 160–61
  on community and immunity, 17, 127–28, 131
  immune response and containment, 137, 143, 144
  on immunity, 16–17, 129–30, 136
  and the "impersonal," 138, 176n31
  on tolerance, 17–18, 175n30
  See also Agamben, Giorgio; biopolitics; Campbell, Timothy

Esquinca, Bernardo, 24, 129

Estatua de sal, La (Novo), 196n16

estrangement, cognitive, 20, 176n35

ethe, 174n22. *See also* baroque ethos

ethos barroco. See baroque ethos

Eugenesia y racismo en México (Suárez y López Guazo), 173n19

Eugenia (Urzaiz), 98, 176n33, 189n27

eugenics, 7, 8, 99
  Lamarckian, 187n16
  Latin America, 173n19
  Mexico and Brazil, 188n25
  and Urzaiz's *Eugenia*, 189n27
  and utopian literature, Brazil, 176n31, 178n25
  See also degeneration; positivism

exception, state of, 131. *See also* Agamben, Giorgio; *Homo Sacer: Sovereign Power and Bare Life* (Agamben)

exceptionalism, 4, 172n13

Expedición a la ciencia ficción mexicana (López Castro), 176n34

Experimental Novel, The (Zola), 173n17

export economy. *See* economics

Eye of Newt and Toe of Frog (Crump), 189n28

fairy tale, 86. *See also* Sleeping Beauty

fantastic, 89–90
  Brazilian authors of, 32, 76–77, 91, 130, 132, 185n5, 190n6, 214n10
  exotic marvelous, 186n14
  Hoffmann, 32
  Machado de Assis, 193n1
  Mexican authors of, 19, 83, 88, 100, 133, 137, 146, 195n12
  Todorovian, 101, 186n14
  See also uncanny

Fantastic, The (Todorov), 186n14

fantasy, 1, 20, 52–53, 70, 176n13
  as export reverie, 113
  and gender, 184n5
  queer, 106
  of reverse colonization, 186n14
  and speculative fiction, 20
  and women, 75, 98, 185n5
  See also fantastic

Fantasy and Imagination in Mexican Narrative (Larson), 21, 176n32

Fausto, Boris, 191n7

favelas, 133, 171n10

federales, 117

feedback loop, 22, 30, 39, 41
  and cybernetics systems, 40, 42, 44, 48, 52
  homeostasis, 30, 43
  and sexuality, 42, 44, 45, 47
  thermodynamics, 39–41, 179n14, 180n17

femicide, 169n1, 181n27, 191n9. *See also* violence

feminism, 28, 30, 74, 102, 187n20, 189n27
  corporeal, 187n20

feminism (*continued*)
 French, 187–88n20
 Haraway, 13, 30
 and proto-feminism, 45, 98, 180n19
 and science fiction, 74, 75, 98
femme fatale, 91, 99–100. *See also* phallic female
Femmenism and the Mexican Woman Intellectual (Hind), 43
Fernández, Bernardo, 23, 56. *See also* Bef
Fernandez, Cid, 180n20
Fernández Delgado, Miguel Ángel, 19, 176n32, 176n34
Fernández Retamar, Roberto, 108
Ficção científica, fantasia e horror no Brasil, 1875–1950 (Causo), 19, 34, 176, 186n9, 186n11
Fideli, Finisia, 189n29
Fifth Empire, 171n12
Figueiredo, André de, 184n1
Filha do inca, A (Del Picchia), 186n10, 188n25
Fischer, Brodwyn, 36
Fischer-Hornung, Dorothea, 141
Fitz, Earl, 173n14
Flaubert, Gustave, 186n11
"Flor, telefone, moça" (Drummond de Andrade), 126
Fonseca, Rubem, 67
Fonsecanazarthe, 195n12
Fordism, 125
Forjando patria (Gamio), 172n13
Foucault, 3, 169n1
 and biopolitics, 17
 and discipline of the body, 179n16
 and sexuality, 6, 14–15, 42–43
 and sovereign power, 170n5
 See also biopolitics
Foundational Fictions, 6
foundational myths, 37, 52, 98, 105, 106, 172n13, 187n15
 and *mestizaje/mestiçagem*, 70–71

Franco, Itamar, 54
Franco, Jean, 187n15
Franco, Marielle, 180n22
Frankenstein, 37, 92, 179n12, 179n15
French, William E., 15
Freud, Sigmund, 30, 150, 177n6, 178n6. *See also* doubles; fantastic; uncanny
Freyre, Gilberto, 9, 156, 172, 174n21, 190n6. See also *mestizaje/mestiçagem*
From Amazons to Zombies, 141, 164
Fuego para los dioses (Pascal), 184n5
Fuentes, Carlos, 19, 24, 136, 144, 146
"Furias de Menlo Park, Las" (Padilla), 181n27
Furlanetto, Elton, 103

Gamio, Miguel, 172n13
Gandler, Stefan, 28, 174n22, 174n25, 175n26, 177nn3–4, 179n13
García, Hernán, 55, 182n32
Garro, Elena, 43–44
Gaslight, 147
Gaspari, Elio, 191n13
gays, 49
 and autobiography, 184n1
 in fiction, 51, 72, 94, 106, 149, 152, 188n23
 in history, 14–15, 187n14, 196n16
 and North American movements, 72–73, 188n22
 and pride marches, 188n22
 and violence in Brazil, 180n22
 See also homosexuality; lesbianism; queerness
Gel azul (Bef), 57–60, 68, 183n35
gender, 2, 8
 and anxiety, 15, 60, 96
 and cyborgs, 29, 31, 36, 45–46, 48–49, 57–60
 in fantasy, 70, 78–79, 88–89, 92–95, 96–97, 99

gender (*continued*)
   fluidity, 90, 105, 188n20
   performativity, 23, 83, 85, 102
   roles, 13, 29, 103, 173n14
   in science fiction, 74–75, 82–83
   and undead, 122–23, 140–41, 149–52,
     161–63, 191n9
   See also Butler, Judith; homosexu-
     ality; homosocial; queerness
Gender and Sexuality in Latin American
   Horror Cinema (Subero), 141
Gender Trouble (Butler), 92
Generation Zombie (Boluk and Lenz), 111
German Reich, 186n13
Ghost in the Shell, The, 62
Gibson, William, 55
Ginway, 19, 33, 48, 176n32, 180n20,
   182n32, 185n5, 186n10, 188n25,
   189n29, 190n6, 193n1
Girard, René, 110
Gledson, John, 178n8
González, Emiliano, 22, 51–53, 181n26,
   181n28
Gordon, Joan, 186n11
Gordus, Andrew, 193n2
gothic, 20, 24–25, 32, 35, 67, 181n26
   in Latin America, 20, 34, 140–41,
     193n6, 195n12
   mansion, 136, 137, 139–40, 146
   monster, 143–44
   novels, 150, 152
   in science fiction, 34, 35, 178n9
   Victorian, 89, 140, 194n9
   See also horror; vampires
Gracos, 161
Graham, Douglas H., 5, 39, 116, 128
Grande arte, A (Fonseca), 67
Great Depression, 112
Green James N., 14
Grosz, Elizabeth, 89, 187n20
Guerrero Heredia, Marco Vladimir, 123
Guimarães, Bernardo, 195n13
Gutiérrez, Gerardo, 171n8

Gutiérrez-Mouat, Ricardo, 144–45
gynocracy, 79, 188n25

hacking, 55–58, 182n32
Halberstam, Judith, 23, 70, 139, 142–44
Haraway, Donna, 13, 44
   and animals, 45–46
   "Cyborg Manifesto," 30–31, 62
   postgender cyborg, 63
   on shared immunity, 162
   tight couplings, 44
Harpold, Terry, 116, 190n5
Haunting, The, 148, 194n8
Haunting of Hill House, The (Jackson),
   148, 194n8
Hausmann, 35
Hayles, Katherine, 13, 177n2
   on cyborgs, 22
   first stage of cybernetics (homeo-
     stasis), 29–30, 33, 39–40, 44, 179n14,
     180n17
   second stage (reflexivity), 44, 52, 61
   third stage of cybernetics (digital),
     53, 55–56, 59, 64–66
   See also cybernetics; reflexivity;
     signifier
Haywood Ferreira, Rachel, 8, 18, 21,
   178–79nn11–12, 187n16, 189n27
Heart of Darkness (Conrad), 96
Hello, Hello Brazil: Popular Music and
   the Making of Modern Brazil
   (McCann), 180n21
"Hemoglobitas, Los" (Cardona Peña),
   193n2
Herbert, Daniel, 184n1
"Herencia de Cthulhu, La" (González),
   181n28
heroic narratives, 23, 50–51, 70, 82–83, 105
Hershfield, Joanne, 39
Hertzog, Vladimir, 132
Hess, David, 94
heterogeneity, 9. *See also mestizaje/*
   mestiçagem

heteronormativity, 6, 14, 35, 60, 82, 85, 90, 95, 96, 180n20. *See also* gender; homosexuality; sexuality: nonnormative
Hewlett and Weinert, 39
Hind, Emily, 43
"Histórias de gente alegre" (do Rio), 188n23
Histórias sem data (Machado), 76
historical novels, 173n15
history, anti-positivist, 181–82n28. *See also* economics
History of Sexuality (Foucault), 14–15, 43, 179n16
Hitchcock, Alfred, 110
Hoeg, Jerry, 31
Hoffmann, E. T. A., 32, 177n6
Holocaust on Trial, 186n13
Holston, James, 119, 173n16
homeostasis, 30, 40, 43, 177n2. *See also* cybernetics; thermodynamics
Homo Sacer: Sovereign Power and Bare Life (Agamben), 17–18
homosexuality, 14–15
    in fiction, 49, 51, 72, 73, 94, 151–52, 162
    and Napoleonic code, 14, 87
    in naturalist texts, 188n23
    outing of, 205n10
    and repression, 91, 175n28
    See also gays; gender; heternormativity; lesbianism; sexuality
homosocial, 23
    male bond, 87–88, 91
    male spaces, 56, 105
    in *Querens*, 37, 179n12
Honores, Elton, 141
Hook, Derek William, 65
"Hora de luciérnagas, La" (Chacek), 192n16
Hora dos ruminantes, A (Veiga), 192n16
horror, 1, 20
    and abjection, 91, 102, 183n33
    in cyborg stories, 32, 36, 51

horror (*continued*)
    in vampire tales, 140–43, 152, 163–64
    in zombie narratives, 110, 118, 124, 131, 133, 191n12
    See also gothic; Kristeva, Julia; monsters
Hour of Eugenics, The (Stepan), 7, 173n19
How We Became Posthuman (Hayles), 177n2
"Huésped, El" (Dávila), 136, 146–47
Human Rights Watch, 180n22
human trafficking, 61, 120–21, 191n9
Humbolt, Alexander von, 81
Hunger, The, 152, 195n11, 195n14
hybridity, 9
    institutions, 28
    and science fiction genres, 81, 176n35
    species, 98, 158, 186n11
    in vampires, 138, 161, 196n15
    See also third way
hyperinflation, 54
hypnotism, 34, 37–38, 178n11
hysteria, 147, 150, 194n9

imaginación: La loca de la casa, La (Bef, ed.), 185n6
immigration
    and the *bracero* program, 159
    Brazil, 66, 164
    European Mexico-Brazil, 7, 170n2
    Japan-Brazil, 196n18
    Mexico-US, 54, 158–59, 161
Immunitas: The Projection and Negation of Life (Esposito), 16
immunity, 16–18
    and auto-immune disease, 17, 127, 150
    and body politic, 16, 134, 144
    and counter-immunity, 18, 108–9, 128, 134, 152
    and governments, 16, 121, 159
    and lack of, 114, 120, 150
    and tolerance, 17, 128, 175n30
    and undead, 17, 161, 196n15

immunity (*continued*)
  See also biopolitics; Esposito,
    Roberto
impersonal, 138, 176n31, 193n3
implants, 183n34. *See also* cyborgs
import substitution, 39. *See also*
    economics
Improper Life (Campbell), 123
Incidente em Antares, 129–32, 191–92n14
income distribution, 112
Inconfidência mineira, 33
indigeneity, 9
indigenous, 3–4, 6–7, 14, 37–38, 67, 71
  in Argentina, 170n2
  assimilationist views, 7, 14, 172n13
  fictional portrayals of, 83–84, 86, 148,
    186n10, 188n25
  goddesses, 170n3
  mothers, 148, 170n3
  vampires, 155, 160
  See also mestizaje/mestiçagem
industrialization, 128, 170n4. *See also*
    economics
inoculation, 128
  against arcane arts, 94
  against bad blood, 152, 154
  against power, 152, 157
  against queer, 87
  See also immunity; vaccination;
    vampires
insecurity, 171n10
Institutional Act No. 5, 129
institutions, 3, 162, 193n2
  hybrid, 28
  of power, 17, 42–43
  and PRI, 118, 128, 188n25
  and slavery, 169n2
  and violence, 119, 181n25
Invasion of the Body Snatchers, The, 109
Iracema (Alencar), 170n3
Islam, Gazi, 28, 74, 103, 106
Island of Dr. Moreau, The (Wells), 80,
    82, 186n9
I Walked with a Zombie, 109

Jackson, Rosemary, 90–91
Jackson, Shirley, 148, 194n8
Jack the Ripper, 155–56
Jaf, Ivan, 157, 195n12
James, Henry, 139
Jameson, Fredric, 53–54
Jane Eyre (Brontë), 194n9
Janzen, Rebecca, 145–46
Japan
  and community in Brazil, 62
  and culture, 196n18
  and cyberpunk, 62–63
  and cyborgs, 62–63, 66
  and manga, 62, 177n1
  and vampires, 164–65, 196n18
Jaula de la melancolía, La (Bartra),
    172n13
Jáuregui, Carlos, 185n8
Jones, Julie, 35
Jorge, Miguel, 192n16
Jrade, Cathy, 141
Juárez, Benito, 7, 100, 144, 172n12
justice killings, 63, 108

Kafka, Franz, 101
Kalum (Del Picchia), 188n25
Kamen, Deborah, 84
Kaori: Perfume de vampira, 138, 162,
    164–65
Kaori e o samurai sem braço (Moon),
    196n17
Kaori 2: Coração de vampira (Moon),
    196n17
Kara e Kmam (Fonseca), 195n13
Kardec, Allan, 94
Keats, John, 139
Khnopff, Fernand, 91
King, Ed, 68
Kittelson, Roger, 174n20
Klor de Alva, Jorge, 9
Krause, James R., 177n6, 192n17
Kristeva, Julia, 50, 60, 86, 91
Kuhnheim, Jill, 83, 86
kyuketsuki (Japanese vampires), 196n18

Labalestier, Justine, 74
Laberinto de la soledad, El (Paz), 170n3, 172n13
"Laboratorio de los espíritus, El" (Becerra), 125
Labouchère Amendment, 175n28
Lacan, Jacques, 64–65
Lacanian theory, 96, 124
La Malinche. *See* Malinche, La
"Lamia" (Keats), 139
Lampert-Weissig, Lisa, 193n4
Langford, David, 176n32
Larson, Ross, 21, 147, 176n32
Latham, Rob, 74, 195n14
Latin American Gothic, The (Casanova-Vizcaíno and Ordiz), 20, 141, 176n32
Latin American science fiction, 1–2, 8, 18-21, 30, 80, 176n32, 179n12
    and Anglo-American, 74, 98
    and cyberpunk, 55–56, 62, 180n20
    and gender and sexuality, 184n5, 186n14
    and mainstream, 50, 65, 67
    and realist canon, 173n15
    and reverse colonization, 154, 160, 186–87n14
    and scientific theories, 186n16
    and women writers, 75
    See also speculative fiction
Latin American Science Fiction: An A to Z Guide (Lockhart), 176n32
Lauro, Sarah Juliet, 110–11
LBGTQ movements, 71, 180n22, 188n22. *See also* gender; homosexuality; lesbianism; queerness
Lear, John, 117
Left Hand of Darkness, The (Le Guin), 74
Legacies of Race (Bailey), 180n21
Le Guin, Ursula, 74, 75
Lehnen, Leila, 119
Lenz, Wiley, 110, 111

Leones, André de, 24, 120
lesbianism
    in Brazilian works, 49, 93, 148, 150–52, 188n23
    and *Haunting of Hill House*, 194n8
    and Latin America, 188n22
    and Mexican works, 153–54
    and pride movements, 188n22
    and *Rebecca*, 151
    See also gender; queerness
Lewis, Colin, 54
Lezama Lima, José, 72
liberalism
    decline of, 170–71n7
    economic doctrines, 190n2
Life between Two Deaths, 1989–2001 (Wegner), 195n14
Lima Barreto, Afonso Henriques, 24, 113, 114–17, 190n4
Literary and Cultural Relations between Brazil and Mexico (Moreira), 173n14
living dead, 2, 8, 24, 107–8, 130–31, 167–68
    as consumers, 123–27
    definitions of, 109–12, 190–91n1
    as economic victims, 116–17
    and immunity, 16, 18, 108–9, 128
    and social classes, 108–9
    as symptoms of crisis, 19
    and trauma, 120–23
    See also vampires; zombies
Living Dead: A Study of the Vampire in Romantic Literature, 139
Living with Insecurity in a Brazilian Favela (Penglase), 181n25
livro vermelho dos vampiros, O (Moon), 195n13
Locating Science Fiction (Milner), 20
Lockhart, Dale, 161, 176n32
Lodi-Ribeiro, Gerson, 25, 137, 154–59, 160, 185n5

Loera, Joaquín Guzmán (El Chapo), 122
logocentrism, 60, 71, 78
Lombroso, Cesare, 143
Lopes, Denilson, 73
López, Rick, 174n20
López Castro, Ramón, 161
López-Lozano, Miguel, 21, 54
Los que moran en las sombras: Asedios al
    vampire en la narrativa peruana
    (Honores), 141
Lost Boys, 195n14
lost civilizations, 81–82, 95n12
lost decade, Brazil, 54
Loudis, Jessica, 191n10
Love, Joseph, 190n2
Lovecraft, H. P., 51, 181n28
Lowe, Elizabeth, 183n27
lucha libre, 40, 174n20, 192n16
Luckhurst, Roger, 176n35
Lumen (Flammarion), 178n11
Lund, Joshua, 14
Lutteroth, Samuel, 174n20
Luzia-homem (Olímpio), 188n23

Machado de Assis, Joaquim Maria
    gendered cyborgs, 22, 37–34, 36
    gothic stories, 178n9, 193n1
    Hoffmann, 177–78n6
    novels, 178n8, 189n1
    race, 36, 178n10
    slavery, 177n7, 178n8
    vampires, 136, 193n1
    women warriors, 23, 76–79, 86, 105
Machen, Arthur, 51
Macías González, Victor W., 14, 175n29
Macunaíma (Andrade), 186–87n14
Madero, Francisco, 100, 117
Madrid, Miguel de la, 54
Madwoman in the Attic, The (Gilbert
    and Gubar), 194n9
Magalhães Júnior, Raimundo, 178n9
magical realism, 19, 174–75n25

Magic Island, The (Seabrook), 109
magnetism, 36, 37, 87, 89, 188n11
mesmerism, 8
somnambulist zombies, 126, 134
See also hypnotism
Mahoney, Phillip, 110, 124
malandragem
    DaMatta, 172n13
    gay movement, 79
    trickster, 186n14
Malinche, La, 51, 86, 170n3
    and history, 187n15
    and Octavio Paz, 172n13
malinchismo, 187n15
"Manifesto antropófago" (Andrade), 79,
    172n13
maquiladoras, 169n1, 181n27
"Marcha das utopias, A" (Andrade), 79,
    185n8
Martínez Morales, José Luis, 193n2
martyrdom, 4, 52–53, 171–72n12
masculinity
    and body, 91, 96–97, 188n25
    heroic, 23, 50
    and nation building, 70–71, 90, 188n25
    and performativity, 59, 75–76, 175n29,
        188n23
    and queering, 84, 87
    and women warriors, 77
    See also homosocial
Matangrano, Bruno Anselmi, 103
maternity
    alternatives to, 100
    and cyborgs, 45, 61, 79, 180n20
    See also motherhood; oviparity;
        reproduction
matriarchy, 79, 185n8. *See also*
    gynocracy
Mattos, Gregório de, 191n11
Maturana, Humberto, 180n18
Maximilian, Emperor, 100, 144
Mbembe, Achille, 2, 169n1

McCann, Bryan, 180n21,
McKee Irwin, 175n29, 187n17
McLuhan, Marshall, 58, 183nn35–36
Meireles da Silva, Alexander, 34, 178n9
Mejicanos en el espacio, (Olvera), 19
*melhores contos brasileiros de ficção cientí-
  fica: Fronteiras, Os* (Causo), 114
Melo, Alfredo César, 9
Memórias póstumas de Brás Cubas
  (Machado), 189n1
Méndez, 194n7
Mercosur, 55. *See also* trade agreements
Merrim, Stephanie, 83
messianism, 4, 171n12
mestizaje/mestiçagem, 2–3, 6, 8–9, 12, 14,
    15, 18, 38
  in Brazil, 156–57
  cybermestizaje, 31, 39, 60–61
  and demythification of, 68, 71, 106
  and foundational fictions, 6, 170n3,
    172n13
  and immunity, 18
  and nationalism, 86, 172n13, 174n20
  and vampires, 138
  See also Black; indigenous; race
Mestizo Modernity (Dalton), 21
Mestizo Nations (De Castro), 104
Metamorphoses (Ovid), 84
Metamorphosis, The (Kafka), 101
Meter, Alejandro, 170n2
Mexican Masculinities (McKee Irwin),
    175n29
Mexican Revolution, 99–100. *See also*
    revolts; revolution
Mexican state, 1950s, 146
Michoacán, 122
Miéville, China, 176n35
Miliotes, Diana, 117
military, 47, 112, 129–30
  comparison in Brazil and Mexico,
    128, 171n7
  and fiction, 45–46, 48–49, 61–62, 96,
    130, 132–33, 160, 193n2

military (*continued*)
  See also authoritarianism;
    dictatorship
militias, 181n25
Milner, Andrew, 20, 176n35
mining, 33, 112
Miravete, Gabriela Damián, 75
miscarriage, 102. *See also* maternity;
    oviparity; reproduction
miscegenation, 6
  in England, 140, 156
  in US, 14
  See also mestizaje/mestiçagem
Mi tío Juan (Urquizo), 188n25
"Mi velorio" (Ripoll Vicente), 190n1
modernism, 89, 187nn16–17, 188n25
modernity
  and consumer society, 124, 127
  and cyborgs, 22, 34–35, 51
  and *ethos barroco*, 10–11
  and Latin America, 108
  matriarchal alternatives, 79
  outsiders to, 36, 38, 67
  resistance, 43–44, 52–53
  See also modernization
modernization, 2, 8, 56
  and cyborgs, 22, 27
  and equality, 5
  and export economy, 29
  and *mestizaje/mestiçagem*, 8
  and positivism, 7
  See also economics; modernity
Molina-Gavilán, Yolanda, 19, 176n32
Monsiváis, Carlos, 196n16
monsters
  and anxiety, 107, 143–44
  and cyborgs, 50, 56, 60, 63
  in Latin America, 20, 21, 24, 62, 108–9
  undying, 122, 137–38, 158, 195n12
  as vampires, 137–43
  as zombies, 108, 110, 127
monstrous
  body, 63, 137, 139

monstrous (*continued*)
  feminine, 60, 63, 99, 140, 195n11
  Frankenstein, 92
monstrum, 20
Monstruo como máquina de guerra, El
  (Moraña), 21, 50, 127, 141
Monteiro, Jeronymo, 19, 81
Monteiro, Luciana, 45, 180n19
Monteiro Lobato, José Bento, 24, 115–17
Moon, Giulia, 25, 138, 162, 164–65,
  196nn17–18
Moraña, Mabel, 21, 50, 56, 60, 127, 141
Moreira, Paulo, 9, 172n14
Moreira-Almeida, Alexander, 94, 188n24
Moreland, Sean, 110
Moreman, Christopher M., 111
Moretti, Franco, 107, 140, 188n21
motherhood, 100, 102, 106. *See*
  also gynocracy; maternity;
  reproduction
mourning, 67, 106, 181n27
  and dictatorship, 50–51
  memorialization, 65
Moussong, Lazlo, 193n2
Mueller, Monika, 141
mujer varonil, 83
Muñoz Fernández, Ángel, 142
Muñoz Trujillo, Gabriel, 25, 75, 138, 154,
  158–61, 176n34, 179n12
Muñoz Zapata, Juan, 55, 182n32
muralists, 118, 173n20
Murilo de Carvalho, José, 171n12
music
  and artistic communities, 95, 180n21
  and censorship, 132
  and nationalism, 8, 39–40, 174n20
mutation
  and baroque, 68
  in cybernetics, 65–66
  and zombies, 111
myths, national, 4. *See also* nation
  building

NAFTA, 5, 175n27. *See also*
  neoliberalism
Napoleon III, 144
narcotrafficking, 6, 55–56, 61, 122
  and cyberpunk, 182n32
  narcogótico, 123
  violence against women, 191n9
  and zombies, 120–23
Nascimento, Abdias do, 10
nationalism, 6, 40, 173n20, 194n7
nationality
  and body, 6
  definitions of, 172n13, 194n10
  and exceptionalism, 4
  myths of, 4, 71, 138, 141, 152
  and pride, 39
  and unity, 6
nation building, 4, 9, 15
  and failure, 67, 83, 90, 106, 172n12
  masculinist discourse, 70, 86
  and *mestizaje/mestiçagem*, 9, 105, 138,
    170n2
  modernization, 38, 53
  and resistance, 72, 90
  state industry, 24
naturalism, 6, 51, 173n15
necropolitics, 2, 18, 111, 119, 167, 169n1
  alternatives to, 128, 138, 148, 176n31
  and disposable bodies, 169n1
  See also biopolitics
Nemi Neto, João, 15, 23, 71, 73, 184n1,
  184nn3–4
neo-baroque, 72, 184n3
neoliberalism, 5, 21, 51, 54, 169n1
  in Brazil, 165, 181n25
  and criminalization of poverty, 182n30
  and cyberpunk, 22, 53, 54–57, 60, 61,
    68, 177n5
  and dismantling of the state, 182n30
  and individualism, 23
  and vampires, 165, 195n12
  and zombie literature, 24, 68, 119

Nervo, Amado, 23, 88–91, 105, 179n15, 187n16
Netflix, 167
neurasthenia, 94
Neuromancer (Gibson), 55
Nevárez, Lisa A, 193n4
Nicholls, Peter, 176n32
Niemeyer, Oscar, 174n20
Night of the Living Dead, The, 110
"Niña de Cambridge, La" (Cardona Peña), 46
Nisei, 164
Noche mexicana, La, 173–74n20
Noche que asolaron Tokio, La (Velázquez Bettancourt), 118
Noffsinger, Robert, 183n36
"No perdura" (Pacheco), 193n2
Nordau, Max, 143
"Nova Califórnia, A" (Lima Barreto), 113–15
novela de la revolución, la, 173n15
Novo, Salvador, 196n16
"Novo protótipo, O" (Causo), 62
novum, 20

Obregón, Álvaro, 117
occult
    in El donador de almas, 88–91
    in Esfinge, 91–95
    practices, 23, 100
    and queering, 87, 95
    sciences, 87–88
    See also pseudo-sciences
Oedipal family, 83, 95
Oedipus, 57
Olvera, Carlos, 19
Olympics, Mexico, 129
Only Happy Ending to a Love Story is an Accident, The (Cuenca), 183n37
"Options" (Varley), 75
O que faz brasil, Brazil (DaMatta), 14
Orbaugh, Sharalyn, 62

Ordem do dia, A (Souza), 193n2
Ordiz, Inés, 20, 141
organic state, 170n6. See also corporate state
organized crime, 2, 5, 182nn30–32, 191n9
    in Ciudad de zombis, 102–23
    in cyberpunk, 55–56, 61–63, 68
    in Gel azul, 59, 183n36
    See also crime; violence
organ trafficking, 62, 68
Orgasmógrafo, El (Serna), 185n5
Origins of the Literary Vampire, The, 139
Orozco, José Clemente, 118
Orozco, Pascual, 117
Ortiz, Fernando, 9, 13
Ortiz de Montellano, Bernardo, 24, 124
"Otra noche de Tlatelolco, La," 129
Outango, 109
Outros mundos (Lodi-Ribeiro), 154
Ovid, 84
oviparity, 100–102, 161, 189n29
oviparous hermaphrodite bats, 159–61
Ovo apunhalado, O (Abreu), 50

Pacheco, Emilio, 193n2
Page, Joanna, 68
Paiva, Marcelo, 191n8
Palmer, Pauline, 150
PAN (Partido Autónoma Nacional), 60
Panopticon, 63
"Pantano de los peces esqueletos, Los" (Avilés), 185n5
Par: Uma novela amazônica, O (Causo), 95
Paraguayan War, 33
parallel state, 55
Partido Autónomo National (PAN), 60
Partido Nacional Revolucionario (PNR), 118
Partido Revolutionario Institucional (PRI), 118
Pascal, H., 184n5

paternalism, 158, 182n30
patriarchal society, 12, 28, 78, 166,
    172n13
    Catholic Church, 137, 146, 154
    challenges to, 83, 90, 100, 105
    private property, work, 82
    resistance to, 97–98, 146–154
pattern recognition, 64, 177n2. *See also*
    cybernetics: third stage
Pau-Brasil à antropofagia e às utopias, Do
    (Andrade), 175n27, 185n8
Payne, Judith, 173n14
Paz, Octavio, 145, 170n3, 172n13
Pearson, Wendy Gay, 185n7
Pedro Páramo (Rulfo), 145
Pellegrini, Tânia, 130
Penglase, Ben, 171n10, 181n25
Pereira, Carla Cristina (pseud., Gerson
    Lodi-Ribeiro), 185n5
Pérez, Genaro, 145
Pérez, H., 49
Perfil del hombre y de la cultura en México
    (Ramos), 172n12
Perlonghernéstor, 72
Peronism, 170n2
Petúnia" (Rubião), 194n10
phallic female, 100, 189n27. *See also*
    Eugenia (Urzaiz); femme fatale
Picatto Pablo, 14, 15
Picchia, Menotti Del, 80, 186n10,
    188n24
"Picnic on Nearside," 75
Picture of Dorian Gray, The (Wilde),
    139, 194n10
Pilettinelson, 112, 132
Pindorama, 185n8. *See also* Andrade,
    Oswald de; "Manifesto antro-
    pófago" (Andrade); matriarchy;
    utopia
"Pirotécnico Zacarias, O" (Rubião),
    190n6
Pirott-Quintero, Laura, 85, 187n15

Pizarro, Francisco, 81
Poe, Edgar Allan, 139
political opening, Brazil, 133
Pons, Maria Cristina, 85
Porcayo, Gerardo Horacio, 55, 183n34
Portinari, Cândido, 174n20
Posada, José Guadalupe, 117, 175n29
Posa Guinea, Rosa María, 188n22
positivism, 6–8
    differences Mexico and Brazil, 173n18
    as ethos, 167
    and military in Brazil, 171n7
Posthumanism and the Graphic Novel
    (King and Page), 68
posthumans, 13, 52, 177n2
    and body, 50, 56, 179n14
    as cyborgs, 29–30, 50, 59, 63, 65–66
    and *mestizaje/mestiçagem*, 68, 69
    and stages of cybernetics, 29–30
    and zombies, 110–11, 191n12
postrevolutionary Mexico, 99–100, 145,
    170–71n7, 172n13
    and corruption, 145–46
    and labor, 128
    and *mestizaje*, 8–9, 172n13
    and military, 171
    and muralists, 118
    and popular culture, 173–74n20
    in Urzaiz's *Eugenia*, 189n27
Potter, Sara Anne, 57, 183n35
poverty, 5, 36, 127, 182n30
Powers of Horror, The (Kristeva), 183n33
Practice of Everyday Life, The (Certeau), 3
Prado, Antonio Anoni, 190n4
Prado, Eugenia, 177n5
Prado, Paulo, 172n13
pre-Colombian civilizations, 81
pre-modernity, 174–75n25, 178n9
Precarious Life (Butler), 166
pregnancy, 58, 78–79, 105, 149. *See*
    also motherhood; oviparity;
    reproduction

Presidente negro, O (Lobato), 176n33
PRI (Partido Revolucionario Institucional), 118
Price, Brian L., 52, 171–72n12
primary resources, 60. *See also* economics
Primera calle de la soledad, La (Porcayo), 55, 182n32, 183n34
"Primera comunión" (Rábago Palafox), 153
privatization, 55, 182n30. *See also* neoliberalism
PRN (Partido Revolucionario Nacional), 118
profanation, 24, 134, 191n12. *See also* Agamben, Giorgio
proto-feminism, 45, 98. *See also* feminism
pseudo-sciences, 6–8, 94
    criminology, 143
    Lamarckian eugenics, 187n16
    occult, 87, 187n16
    racism, 36, 115, 178n10
    spiritism, 94–95, 178–79nn11–12, 188n24
    Theosophy, 87, 187n16
    See also hypnotism; magnetism
Pygmalion, 42, 179n15

"Quando é preciso ser homem" (Fideli), 185n5
Quarantine, 129, 130
Queer Art of Failure (Halberstam), 23, 73
queerness, 15, 23, 28, 87, 106
    anthropophagic queer, 71–73
    in Brazilian literature, 83–85, 95–98, 102–4
    in cyborg narratives, 48–50, 62
    and deconstruction of gender, 78–79
    and gender performativity, 23, 75–76, 85

queerness (*continued*)
    in Mexican literature, 84–91, 99, 102
    and space in-between, 28, 73–74
    See also gender; homosexuality; homosocial; lesbianism; occult; sexuality
Queer Universes: Sexualities and Science Fiction (Pearson ed.), 185n7
Queiroz, Dinah Silveira de, 22
Queluz, 188n25
Querens, 36, 187n16, 178–79n12
Quidquid volueris (Flaubert), 186n11
Quincas Borba (Machado), 178n8

Rábago Palafox, Gabriela, 25, 153
Rabasa, Emilio E., 7
race, 3–4, 38, 109, 111n9, 172n13
    and European immigration, 7
    in literature, 36, 61, 62, 84–85, 95, 148, 157–58, 161
    and *mestizaje/mestiçagem*, 6, 9–10, 38, 138
    and poverty, 5, 18
    racial democracy, 10, 180n21
    See also Black; eugenics; indigenous; *mestizaje/mestiçagem*; slavery
racism, 14, 156, 173n19
Raízes do Brasil (Buarque de Holanda), 172n13
Rama, Ángel, 9, 10
Ramos, Samuel, 172n13
randomness, 65–66
rape, 47, 81, 84, 96–97, 105, 146. *See also* violence
Raza cósmica, La (Vasconcelos), 9–10, 138, 172n13
realism in Mexico and Brazil, 173n15, 180–81n24
realist ethos, 59, 175n25
Rebecca (Du Maurier), 147, 150, 194n9
Rebolledo, Efrén, 23, 98, 189n27

reflexivity, 22, 30, 44, 52
  and cybernetic theory, 177n2, 180n18
  in cyborg stories, 46–53, 61–62, 64–66
  definition of, 30, 44
  and second stage cybernetics, 30,
    52–53
  See also feedback loop; homeostasis
reincarnation, 94. *See also* spiritism
rejuvenation, 82, 125, 186n11
Relações de sangue (Argel), 195n13
reproduction
  asexual, 99–100, 161
  and cloning, 95
  female role in, 50
  and Kristeva, 62
  male, 189n27
  nation building and, 86, 98, 105–6
  queering of, 72, 75, 76
  See also oviparity
reptiles, 23, 76
  and occult, 100, 189n28
  and oviparity, 189n29
  and women, 98, 102
resilience, 2, 167
  and antropofagia, 28
  and baroque ethos, 11–12
  and theories of immunity, 16
  See also resistance
resistance, 10–11
  and body, 20–22, 38, 44–45, 49, 167
  and capitalism, 28–29, 34, 41, 56, 59,
    108, 175n27
  and feminism, 28, 45, 63
  and gender, 15, 23, 74, 83, 97–98,
    102, 105
  and immunity, 16, 18, 114, 127–28,
    134–35, 137, 167
  and vampires, 24–25, 137–38, 144–45,
    152, 156, 159
  and writing, 36, 41, 90
  and zombies, 24, 107–8, 128, 135
Retrato do Brasil (Prado), 172n13

return of the repressed
  definition of, 178n6
  historical memory, 33
  and race, 153
  sexuality, 94, 150–52
  violence, 130–31, 157
revenant, 137, 144. *See also* undead, defi-
    nitions of; zombies
reverse colonization, 140, 156, 160,
    186–87n14
revolts
  in Brazil, 117, 191n7
  Inconfidência Mineira, 33
  See also revolution: Mexican
revolution
  and calaveras, 117–18
  Echeverría, 12, 175nn26–27
  Mexican, 4, 38, 112, 117–18
  novel of, 173n15
  positivist, 7
  utopian, 10, 79, 175n27, 185n8
  Zapatista, 177n4
  See also postrevolutionary Mexico;
    revolts
Reyes, Afonso, 173n14, 174n21
Ribeiro, Darcy, 172n13
Ribeiro, Gerson Lodi. *See* Lodi-Ribeiro,
    Gerson
Rice, Ann, 152
Rieder, John, 186n12
Riley, Brandon, 110
Rio, João do, 22, 35–36, 178nn9–10, 188n23
Ripoll Melo, Juan Vicente, 189–90n1
Rivera, Diego, 118
Rivera, José Eustacio, 55, 112
robots, 26–27
  doll-like, 22, 27, 42, 181n7
  and homosexuality, 48–51
  and reproduction, 49, 180n20
  and sex, 37, 41–42, 44, 63–66, 51–53,
    181n26
Rodrigues, Antônio Edmilson Martins,
    178n9

Rojas Hernández, Arturo, 184n5
Rojo, Pepe, 22, 56, 183n34
Romagnac, Carlos, 15
romance-reportagem, 51, 180–81n24
romanticism, 6. See also foundational
    myths; Sommers, Doris
Romero, George, 110
Rondon, Cândido, 81
"Rosas brancas" (Causo), 61
Rubião, Murilo, 192n6, 194n10
"Rudisbroeck o los autómatas"
    (González), 51–53, 68, 181n26
"Ruido gris" (Rojo), 56, 183n34
"Ruinas circulares, Las" (Borges), 52
Rulfo, Juan, 145
Rüsche, Ana, 103
Rushton, Cory James, 111
Ruta del hielo y la sal, La, 138, 162–64

sacred
    bodies, 170n3
    as "impersonal," 138, 176n31, 193n3
    mother, 3
    and profanation, 24, 123–24, 134–35
    waters, 83
    See also impersonal; profanation;
    salvific
Sailendra, Kalar (pseud., Arturo Rojas
    Hernández), 184
salamander, 99, 189n28
"Salamandra" (Rebelledo), 98–100
Salcedo, Doris, 184n3
Salón de belleza (Bellatín), 184n5
salvific, 163. See also sacred
Sánchez Prado, Ignacio, 9, 169n1,
    182n30
"Sandman, The" (Hoffmann), 31, 177n6
Sangre de la medusa, La (Pacheco),
    193n2
Santiago, Silviano, 188
    and entre-lugar, 73, 184n2

Santiago, Silviano (continued)
    on homosexuality, 72–73
    and space-in-between, 73, 184n2,
        188n20
    on naturalism, 87
    and wiliness, 23, 72–73
Sardà, Alejandra, 188n22
Sarduy, Severo, 72
Sarney, José, 54
scandal of the "41," 175n29
Schaffer, Talia, 193n5
Schlofsky, Sebastián, 169n1
Schwarcz, Lilia Moritz, 173n19
Schwarz, Roberto, 12, 31
science fiction
    and adaptation to Latin America, 8
    and antropofagia, 23, 80, 108, 136
    and body, 1–2, 56, 72, 74, 167
    and mainstream, 65, 67, 114, 176n35
    traditions in Mexico and Brazil, 18–20,
        173n15, 176n32
    and selective tradition, 20
    and speculative fiction, 2, 5, 52, 90,
        110, 168
    See also cyberpunk; dystopia; fantas-
        tic; fantasy; gothic; utopia
Science Fiction: A Very Short Introduction
    (Seed), 176n35
scientific racism, 7, 38, 173n19. See also
    eugenics
Seabrook, William, 109
Sedgwick, Eve Kosofsky, 75, 87–88. See
    also homosocial
Seed, David, 176n35
Seltzer, Mark, 179n14
"Selva de fantasmas" (Eljaiek-
    Rodríguez), 140
"Semana de Arte Moderna, A," 174n20
"Seminário dos ratos" (Telles), 132–33,
    192n17
Serna, Enrique, 185n5
Serra, M. V., 178n9

Serrano, Carmen, 117–19, 135, 141, 194n7
Serrato Córdova, José Eduardo, 53, 181n26
Sete, Os (Vianco), 138, 161, 195n15
Sexual Anarchy: Gender and Culture at the Fin de Siècle (Showalter), 175n28
sexuality
 and anarchy, 15, 50, 70, 87, 90, 94
 and cyborgs, 30, 36, 40, 43, 60
 female, 82, 86, 88–94, 109, 185n5
 nonnormative, 14–15, 23, 48–51, 70, 72–73, 86, 104–5
 in science fiction, 74, 185n7
 and vampires, 9–14, 156, 162–64
 See also homosexuality; lesbianism; queerness
Shannon, Claude, 180n17
Shelley, Mary, 37, 92
Shelley, Percy, 139
Shildrick, Margrit, 63
Shiroma, matadora ciborgue (Causo), 61–63
Showalter, Elaine, 175n28
Siegel, Don, 109
Sierra, Justo, 7
Sifuentes, Gerardo, 56
Sifuentes-Jáuregui, Ben, 15, 75–76, 175n29
signifier
 and Amazon, 80
 flickering, 30, 53, 64–66
 and *mestizaje/mestiçagem*, 8
 split, 97
 and vampires, 141
 and zombies, 111
Silva, César, 196n18
Silva, Chica da, 157–58
Silva, Luiz Ignacio da, 120
"Singular ocorrência" (Machado), 77
Skidmore, Thomas, 7, 54, 112
Skin Shows: Gothic Horror and the Technology of Monsters, 142, 143

skulls
 as *calaveras*, 117–48
 in *Eugenia*, 189n27
 and proto-cyborgs, 32–33, 35
 and slave labor, 34, 67
 and vampires, 143
Slater, Candace, 80
slavery, 2, 4, 67, 104, 105, 172n13, 178nn7–8
 and export economy, 31, 33
 female, 62, 96, 186n11
 in Mexico and Brazil, 169n2, 171n8
 and rural labor, 36, 114, 153
 vampires, 155–57, 160, 195n13
 and zombies, 34, 111, 116
Sleeping Beauty, 52, 83, 85
Smith, Peter H., 54, 112
Snyder, Zack, 110
soccer, 174n20
social body, 2–3, 35
social Darwinism, 7. *See also* positivism; scientific racism
socialism in Mexico, 128, 188n25
Society Must Be Defended (Foucault), 170n5
"Soliloquilo de un muerto" (Bermúdez), 189n1
Sommers, Doris, 6, 8. *See also* foundational myths
somnambulist zombies, 109–10, 126, 134
Souza, Márcio, 193n2
sovereign power, 145, 170n5
space in-between, 73, 94, 184n2. *See also* Santiago, Silviano
speculation and wealth, 182n31
speculative fiction
 definition of, 1
 and genre hybridity, 176n35
 originality of, 167
spiritism, 94, 179n12
 and alternative or pseudo sciences, 187n16

spiritism (*continued*)
    and madness, 188n24
stages of cybernetics, 177n2
state capitalism
    and commodity fetishism, 179n13
    and import substitution, 5, 29, 39, 41,
        54, 111
    and industrialization, 26, 39, 44
    and protectionism, 39, 112
Stepan, Alfredo, 170n6
Stepannancy Leys, 173n19
Stevenson, Robert Louis, 87, 88, 98, 139
Stoker, Bram, 91, 139, 155
*Strange Case of Dr. Jekyll and Mr. Hyde,
    The* (Stevenson), 87, 98, 139
Suárez y López Guazo, Laura Luz, 7,
    173n19
subaltern
    and *bíos*, 18, 137
    and colonization, 12
    and *ethos barroco*, 10–11, 27–28,
        174n24
    and *zoê*, 17–18, 146, 165–66
    and zombie resistance, 18, 28,
        118–19, 128
Subero, Gustavo, 141
Sueños de la bella durmiente (González), 52
supplementarity, 182n29
Süssekind, Flora, 51, 173n15, 180n24
Suvin, Darko, 20
swarms
    as counter-immunological response,
        131–33
    of insects, 192n16
    of rats, 132
    as response to oppression, 192n16
    as zombies, 110
syncretism, 9, 12
    religious, 87, 94, 170n3
synesthesia, 89

Taibo, Paco Ignacio, II, 57
Tal Brasil, qual romance (Süssekind),
    173n15

Tarazona, Diana, 23, 100–101, 106,
    189n29
Tavares, Braulio, 77, 176n32
Tavares, Eneias, 103
Taylor, Claire, 84
techo-fictions, 29. *See also* cyberpunk;
    science fiction; speculative fiction
Telles, Lygia Fagundes, 24, 132, 185n5,
    192n17
Tepes, Vlad, 138, 195n12
testimonial novels, 173n15. *See also* nat-
    uralism; novela de la revolución,
    la; realism in Mexico and Brazil
thanatopolitics. *See* necropolitics
Them, 110
thermodynamics, 39–40
    and first stage of cybernetics, 180n17
    laws of, 179n14
    and motor, 180n17
    See also homeostasis
third phase of cybernetics, 64–65,
    177n2. *See also* signifier: flickering
third way, 28–29, 97–98. *See also*
    hybridity
3% (Aguilera), 167
"Tia Nela" (Serna), 185n5
Tichi, Cecelia, 125
"Tigrela" (Telles), 185n5
Tiptree Jr., James, 98, 188n26
"Tlactocatzine en el jardín de Flandes"
    (Fuentes), 137, 144, 146
Tlatelolco, 47
Todorov, Tzvetan, 101, 186n14
Tola de Habich, Fernando, 142
Tomorrow's Eve (Villiers L'Isle Adam),
    179n15
Toro, Guillermo del, 164
torture, 130, 131
    in Brazil, 191n13
    and cyborgs, 177n5
    and military in Argentina, 170n2
trade agreements, 5, 21, 54, 55, 175n27,
    182n30

transculturation, 9, 10, 13
transgender characters, 184–85n5
transmigration, 23. *See also* spiritism
"Transplante de cérebro" (Carneiro),
 185n5
trans-species, 76, 97, 100, 102, 186n11
transvestism, 49, 75–76, 85, 175n29
trauma
 collective, 107
 and cyborgs, 56, 97
 and dictatorship, 30, 60
 psychological, 95–96, 101–2
 trauma theory, 97
 vampires, 144
 zombies, 24, 107–8, 116, 119–23, 134
Trevisan, Dalton, 193n2
Trevisan, João Silvério, 14, 187n14
Triste fim de Policarpo Quaresma (Lima
 Barreto), 190n4
tropical gothic, 152, 193n6, 194n10
Tropical Gothic in Literature and Culture
 (Edwards and Gardini
 Vasconcelos), 20, 141
Trujillo Muñoz, Gabriel, 25, 137, 154,
 158–61, 176n34, 179n12
28 Days Later (Boyle), 110
Twilight saga, 161
Twitchell, James B., 139
Tyson, Peter, 186n13

"Últimas horas de los últimos días, Las"
 (Bef), 185n5
uncanny
 and Amazon, 80
 and cyborgs, 32, 44, 52, 68
 doubles, 121, 150
 Freudian, 178n6
 supernatural, 93–94, 150
 zombies, 116, 124
undead, definitions of, 189n1. *See also*
 revenant; vampires; zombies
underclass, 4. *See also* subaltern
Understanding Media (McLuhan),
 183n35

*Único final feliz para uma história de
 amor é um acidente, O* (Cuenca),
 63–67
Urzaiz, Eduardo, 98, 176n33, 189n27
utopia, 8, 18, 171n12, 175n26
 and alternatives, 63, 86, 90, 106,
 158, 161
 and Amazon, 79–80
 Aztlán, 9
 conservative, 188n25
 in *Eugenia*, 189n27
 as literary genre, 19
 and literature Brazil and Mexico,
 176n33
 and Pindorama, 185n8
 and tales of redemption, 31, 160
 utopia of a modern nation, 51,
 180n24
Utopian Dreams, Apocalyptic Nightmares
 (López-Lozano), 21

vaccination, 133, 191n7. *See also* immunity; inoculation
Valek, Aline, 23
value
 market, 11, 29, 33, 179n13
 surplus, 119
 use, 45
vampires, 24, 162
 as agent of the oppressed, 154–62,
 164–66
 and degeneration, 136–37, 143
 as exploiter, 136, 142–46, 160, 193n2,
 196n15
 and female sexuality, 25, 137, 140,
 146–52, 195n11
 as human hybrid, 138, 161–62
 and immunological response, 25,
 142, 146
 Latin American literature, 140–41,
 193n20
 and *mestizaje/mestiçagem*, 138,
 158–61, 153
 vegetal, 152, 194n10

"Vampiro, El" (Cuevas), 136, 142

Vampiro, El (Méndez), 141

Vampiro antes de Drácula (Argel and Moura), 195n13

Vampiro da Mata Atlântida, O (Argel), 195n13

Vampiro da Nova Holanda, O (Lodi-Ribeiro), 137, 154

Vampiro de cada um, O (Argel), 195n13

Vampiro de Curitiba, O (Trevisan), 193n20

Vampiro que descobriu o Brasil, O (Jaf), 157, 195n12

Vampiros no espelho (Moon), 195n13

Vampiros podem estar onde você menos imagina (Argel), 195n13

Vargas, Getúlio, 174n20, 180n21

Varley, John, 74

Vasconcelos, José, 8, 9, 172n13, 174n21

Vasconcelos, Sandra, 20, 141

Veiga, José J., 192n16

Velázquez Bettancourt, Diego, 24, 118–19

Veríssimo Erico, 24, 130–31, 191n14

"Véspera de pânico" (Jorge), 192n16

"Viajero, El" (Zárate), 57

Vianco, André, 25, 138, 161, 195n15

Vicente, Gil, 131

"Vida eterna, A" (Machado), 136, 193n1

"Vida possível atrás das barricadas" (Barcia), 180n20

Vieira, Antônio, 171n12

Viking culture, 81. *See lost* civilizations

Villa, Andrea, 184n3

Villa, Pancho, 117

Villaba, Verónica, 188n22

Villa-Lobos, Heitor, 174n20

Villiers L'Isle Adam, Auguste, 179n15

Vint, Sherryl, 119, 191n12

violence, 3, 5–6, 16

    against women, 28, 66, 119, 156–57, 191n9

    and colonization, 10, 12, 23, 67, 81, 105

violence (*continued*)

    and domestic abuse, 148

    and immunity paradigm, 16, 18, 128

    and LBGTQ populations, 48–50, 77–78, 180n22

    and low-income populations, 5–6, 117, 173n16, 182n30

    and neoliberalism, 51, 54–57, 60–61, 68, 120, 121, 181n25, 183n34

    political, 117, 130–32, 177n5

    See also dictatorship; military; organized crime

Violence and the Sacred (Girard), 110

Virgin Mary, 47, 148

Virgin of Guadalupe, 148, 170n3

virus

    in body politic, 16

    and computers, 56

    and consumer capitalism, 24, 108, 113

    work as, 134

    and zombies, 120

    See also immunity

Visiones periféricas (Fernández Delgado), 176n34

Vivanco, José Miguel, 180n22

Vlad (Fuentes), 195n12

Vlad the Impaler, 138, 195n12

Volatile Bodies: Towards a Corporeal Feminism (Grosz), 187n20

Vorágine, La (Rivera), 55, 113

voudou, 120

Voz de sangre, La (Rábago Palafox), 153, 193n2

Wallace, Alfred, 81

War on Drugs, 161

Webb, Jen, 121, 124–25

Wegner, Phil, 195n14

Weil, Simone, 193n3

weird fiction, 176n35

Weird Tales, 51

Wells, H. G., 80, 186n9

werewolf, 155–56, 165
White, Patricia, 148
White Zombie, 109
Wienernorbert, 30, 40, 180n17
Wilde, Oscar, 87, 91, 139, 175n28
Wind-Up Girl, The (Bacigalupi), 66
Wise, Robert, 194n8
witch, 100. *See also* femme fatale; monstrous: feminine; phallic female
women's suffrage, 91
women warriors, 23, 76–86
World War Z (Forster), 110
Wyllys, Jean, 180n22

Xanto: Novelucha libre (Zárate), 192n16
xenophobia, 156
"Xochiquetzal" (Pereira), 185n5
XXyërröddny, donde el gran sueño se enraízan (Sailendra), 184n5

Yanielli, Joseph, 178n7

Zapatistas, 175n27, 177n4
Zárate, José Luis, 25, 57, 138, 162, 164, 192n16, 196n16
Zé do Caixão, 141
zoê, 17–18, 146, 165–66, 169n17
Zola, Émile, 173n17
zombie-*calaveras*, 118–49
zombies, 2, 16–17, 24
  American vs. Latin American typology, 110–11
  consumer, 123–27, 191n12
  and counter immunity, 17, 107–11
  political, 127–34
  as posthuman, 110–11
  and resistance, 108–9
  and speed, 110, 116, 190n5
  and trauma, 111–22
  See also cannibal; immunity; revenant; undead, definitions of; *zoê*
zones of abandonment, 165

CPSIA information can be obtained
at www.ICGtesting.com
Printed in the USA
LVHW050809060121
675679LV00005B/389